B-25 Mitchell

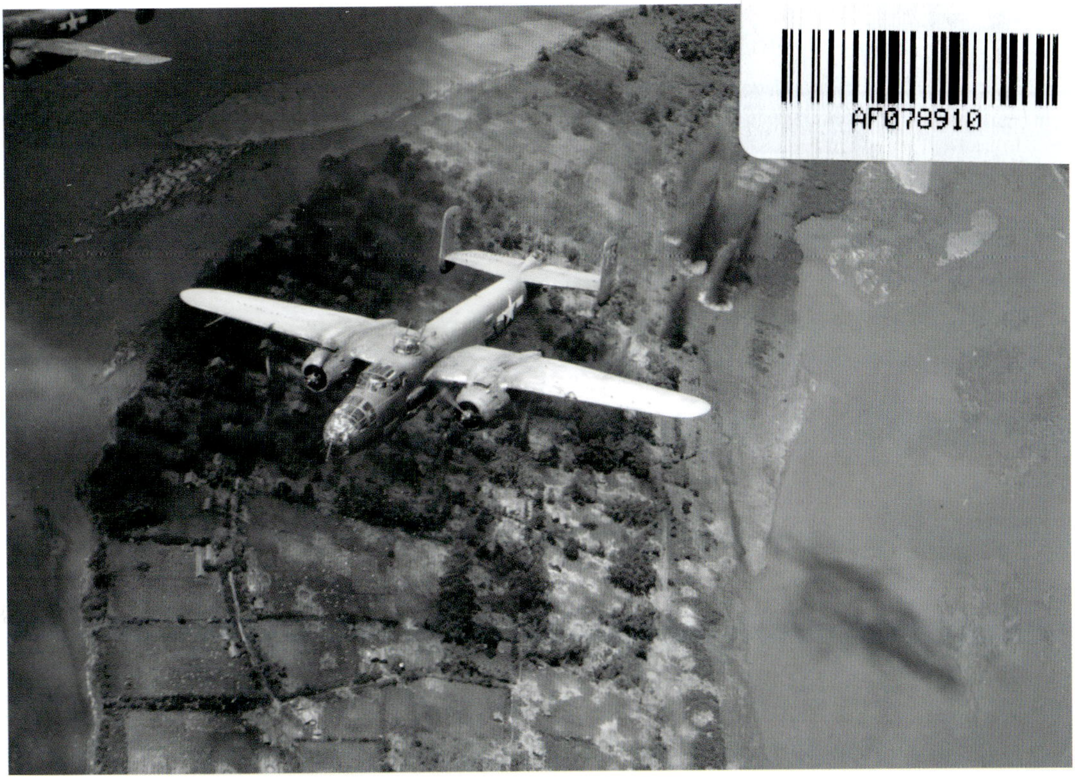

THOMAS McKELVEY CLEAVER

KEY Books

HISTORIC MILITARY AIRCRAFT SERIES, VOLUME 12

Front cover image: This B-25G, 42-64758, was converted from B-25C format, and here it is seen conducting trials with the United States Army Air Forces (USAAF) Technical Center, Orlando, Florida. The "G" was the first Mitchell to field a 75mm gun, this being a nose-mounted weapon. (Key Collection)

Title page image: A B-25J, tailcode 7S, 12th Bomb Group (BG) (probably 434th Bomb Squadron [BS]), leaves behind its bombs strikes on a Burmese target in 1944. (Key Collection)

Contents page image: Mitchell IIs of the Royal Air Force's (RAF) 180 Squadron fly in echelon. The unit was stationed at Foulsham, Norfolk, UK, during 1943. (Key Collection)

Published by Key Books
An imprint of Key Publishing Ltd
PO Box 100
Stamford
Lincs PE19 1XQ

www.keypublishing.com

Original edition published as Combat Machines No. 2
B-25 Mitchell by Key Publishing Ltd © 2017

ISBN 978 1 80282 317 2

All rights reserved. Reproduction in whole or in part in any form whatsoever or by any means is strictly prohibited without the prior permission of the Publisher.

Typeset by SJmagic DESIGN SERVICES, India.

Contents

Chapter 1 Development: Creating a Champion 4

Chapter 2 Tokyo Raid: Roosevelt's Reply 17

Chapter 3 Export Users: Worldwide Mitchells 28

Chapter 4 Gun Noses: The Strafers 41

Chapter 5 The Real *Catch-22*: Battle of the Brenner Pass 60

Chapter 6 Leatherneck Ops: Semper Fi Mitchells 78

Chapter 7 War's End: *Betty's Dream* 92

Chapter 1

Development
Creating a Champion

North American Aviation's (NAA) B-25 was named in honor of General William L. "Billy" Mitchell, Commander of the US Army Air Service on the Western Front during World War One and later a prophet of airpower, who pushed his argument against the perceived wisdom of the day to the point of court-martial, after demonstrating that bombers could sink a battleship.

The Mitchell was the most widely produced American medium bomber of World War Two, serving in every theater of the war and flown by American, British, Soviet, Australian and Dutch airmen. The most common twin-engined aircraft in the postwar US Air Force (USAF), it also served with air arms around the world for a further two decades, and today is one of the most widely seen warbirds on the international airshow circuit. Between 1940 and 1945, a total of 9,889 Mitchells rolled out of NAA factories in Los Angeles, California, and Kansas City, Kansas, where a factory in Fairfax (a suburb of Kansas City) produced 6,683 B-25s, including the overwhelming majority of those still in existence.

On April 18, 1942, the Mitchell was immortalized in the annals of history, when 16 examples took off from the aircraft carrier USS *Hornet* (CV-8) to bomb Japan. Led by legendary airman Lieutenant Colonel James H. "Jimmy" Doolittle, the attacks on Tokyo, Yokohama, and Kobe-Osaka took place just 17 weeks after the Japanese surprise attack on Pearl Harbor catapulted the US into the war.

The B-25B was the first combat-capable version of the Mitchell bomber. Modelers should note how the Olive Drab camouflage faded under sunlight, with the fabric-covered control surfaces wearing lighter than the aluminum airframe. (NARA)

North American Aviation's (NAA) response to a 1938 Air Corps requirement for a twin-engined attack bomber was the NA-40. The prototype is seen in early configuration, without the glass-nose bombardier position. (NARA)

Had the Mitchell done nothing else, its place in history would be secure. However, the bomber set other records around the world in the three-and-half years after Pearl Harbor. Mitchells armed with as many as 14 forward-firing .50 cal machine guns and weapons as large as a 75mm cannon swept the Pacific from Port Moresby to Japan itself. In Italy, B-25s of the 57th Bomb Wing (BW) carried out "The Battle of the Brenner Pass" between November 6, 1944, and April 1, 1945, during the coldest European winter in a century, cutting supplies to German armies in northern Italy and finally bringing the war on the peninsula to an end. The men and aircraft of the 57th entered modern lexicon when a bombardier in the 488th Bomb Squadron (BS), 340th Bomb Group (BG) used his memories of the campaign to write the most famous American war novel to come from World War Two – *Catch-22*.

What ultimately became the B-25 resulted from Circular 38-385, issued by the Army Air Corps (AAC) in March 1938, which described the performance the airmen believed would be required of the next generation of attack bombers. The expectation was a bomber with performance that exceeded single-engined types (such as the Northrop A-17) then in service, and required a payload of 1,200lb (540kg), a range of 1,200 miles (1,900km), and a top speed more than 200mph (320km/h). The primary goal of the new bomber would be battlefield interdiction.

Designed and flown in 1936, the XB-21 was NAA's first attempt at creating a twin-engined medium bomber. The performance of the XB-21, proposed as a replacement for the Douglas B-18, was insufficient to proceed with development past the single prototype. (NARA)

Phoenix-like Origins

North American Aviation was founded on December 6, 1928, as a holding company engaged in the purchase and sale of interests in airlines and aviation-related firms. After such holding companies were broken up by the Air Mail Act of 1934, NAA became an aircraft manufacturing company. James H. "Dutch" Kindelberger, an extrovert aeronautical engineer famed for his emphasis on hard work, orderliness, and punctuality, became president and general manager when he was recruited from Douglas Aircraft, where he had participated in the design and development of the DC-1 and DC-2 airliners. Kindelberger brought with him from Douglas a brilliant young engineer, J. L. "Lee" Atwood, who became chief designer. Kindelberger's first act was to move the company's operations from Dundalk, Maryland, to Inglewood, California, a suburb of Los Angeles, where the weather enabled year-round flying, at a site just south of what was then known as Mines Field; today's Los Angeles International Airport (the NAA factory is now the airport cargo area). He decided to focus on the development of training aircraft, the theory being that this was easier than competing with established companies on larger projects.

The first NAA designs were the NA-15 observation aircraft, which became known as the O-47, and the NA-16 trainer (originally ordered as the BC-1), before being developed into the BT-9, for which NAA received a US$1m production order; eventually, the BT-9 resulted in the AT-6 Texan, the most widely produced trainer after the Soviet Polikarpov Po-2. NAA also designed the NA-39 in 1936, a partially successful medium bomber ordered initially by the AAC as the XB-21, but then canceled due to its disappointing performance.

Responses to Circular 38-385 were submitted by Bell, Boeing-Stearman, Douglas, Martin, and NAA. The Model 9 from Bell was to be an aircraft powered by two Allison in-line engines, but it was withdrawn from competition. Boeing-Stearman's Model X-100 was a three-seat high-winged monoplane with two untried Pratt & Whitney R-2180 radials. Martin's Model 139 was essentially a modernized B-10, by then a six-year-old design. The Douglas Model 7B was a high-winged monoplane powered by two 1,100hp Pratt & Whitney R-1830 Twin Wasp radials.

NAA's contribution, the NA-40, had a five-man crew – pilot, copilot, bombardier/navigator, radio operator/gunner, and gunner – unlike its competitors, which all featured three crewmembers. Pilot and copilot sat in tandem under an expansive canopy, the bombardier/navigator was positioned in a greenhouse-type nose, and the radio operator and gunner were in the aft fuselage. The shoulder-mounted

The three squadrons of the 17th BG were the first to receive the new B-25A Mitchell in September 1941. (NARA)

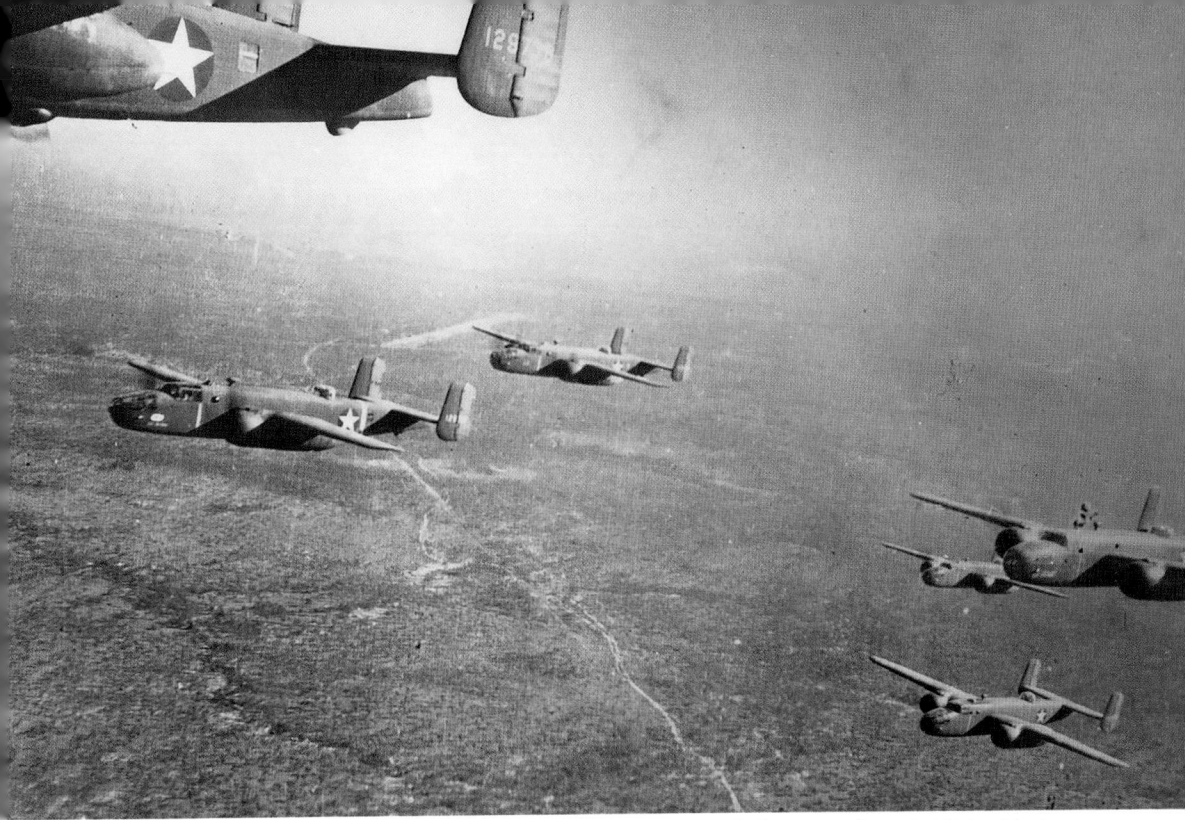

A formation of early production B-25Cs. Crew training for the B-25 Mitchell was conducted at Columbia Army Air Field, South Carolina. (NARA)

cantilever wings had a dihedral of 3° 23', with two Pratt & Whitney R-1830-S6C3-6 air-cooled radials rated at 1,100hp mounted in underwing nacelles. The horizontal stabilizer had two rudders mounted at the tips behind the engines, which would provide controllability in event of engine loss. A tricycle undercarriage was adopted with the nosewheel under the cockpit behind the glass nose, and the mainwheels retracting into the rear of the engine nacelles. Armament was one .30 cal machine gun ball-mounted in the nose, a second in a dorsal turret behind the cockpit, and a third that could be swung between waist and ventral positions. The proposed two fixed machine guns in the wings were not installed.

Solving The Issues

Carrying the civil registration NX14221, the NA-40 first flew on January 29, 1939, with company test pilot Paul Balfour in the pilot's seat. Early tests revealed severe tail shaking, which worsened as speed increased. Engine cooling problems resulted in long tail pipes being fitted to the exhaust stacks, but the problems continued. Maximum speed was 268mph (431km/h) at 5,000ft (1,524m), and the general feeling was that the NA-40 lacked power, which resulted in Wright R-2600-A71-3 Double Cyclone 14-cylinder air-cooled radial engines rated at 1,600hp being installed in February 1939. The prototype was redesignated NA-40B, first flying on March 1, 1939, and demonstrating improved performance with a maximum speed of 287mph (462km/h) at 5,000ft. The aircraft arrived at Wright Field on March 12, 1939. Flight tests revealed few problems, with the only real complaint being that the Curtiss Electric propellers feathered slowly, which delayed engine-out testing.

On April 11, 1939, the pilot lost control during an engine-out test. While the crew escaped without serious injury, the NA-40B caught fire in the crash and was a total loss. Although the basic design was found not at fault, the AAC nevertheless awarded the attack bomber contract to Douglas in July 1939 for a revised version of the Model 7B, the DB-7, designated A-20.

B-25As on the flight line at NAA's Inglewood factory in autumn 1941. (NARA)

While this attack bomber contest continued, the AAC issued a specification for a medium bomber in March 1939; "medium," in this sense, meant a bomber that would operate from medium altitude, i.e. 6,000–14,000ft (1,828–4,267m), and which called for a payload of 2,400lb (1,100kg), a range of 1,200 miles (1,900km), and maximum speed of 300mph (480 km/h). NAA responded with the NA-62, which was ordered "off the drawing board" in September 1939 as the B-25 Mitchell, along with the design submitted by the Glenn L. Martin Company, which became the B-26 Marauder. The B-25 was powered by the same Wright R-2600s fitted to the NA-40B, while Martin's bomber employed the new Pratt & Whitney R-2800. The two aircraft would be seen throughout the war as competitors for performance and the affection of their crews. John W. Young, who flew both types in the 319th BG in 1942–44, recalled the B-26 as being "like a Packard limousine," while the B-25 reminded him of "a tinny Model T." The B-25 never suffered the handling problems the B-26 faced because of its smaller wing, which resulted in the latter aircraft becoming known as "The Baltimore Whore" for its alleged lack of a visible means of support. The B-25 was considered far easier to handle following the loss of an engine; there was never "one a day in Tampa Bay" with the Mitchell, as there was with Martin's Marauder.

The bulbous tail gun position of the first B-25s was an immediate recognition feature. Note the smaller vertical tails and rudders of this early Mitchell. (NARA)

B-25Ds assigned to antisubmarine patrol fly in formation. Note the overall white lower camouflage of these aircraft. (NARA)

The AAC order for 184 Mitchells was delivered in three subtypes, as testing, operational experience and study of European combat demonstrated a need for changes.

The first production model B-25 flew in August 1940. The first nine Mitchells had a constant-dihedral wing, which caused a lack of stability that was resolved by reducing the dihedral to zero degrees outboard of the engine nacelles, which nullified the problem and gave the airplane its distinctive 'gull-wing' configuration. A less-noticeable physical change was an increase in the size of the twin tail fins and a decrease in their inward tilt. The first 24 B-25s had a bulbous tail gun position with clamshell doors, which opened to enable traverse of the single .50 cal machine gun, while there was .30 cal machine gun in the bombardier's nose position and three additional .30 cal weapons firing through hatches in the waist: upper, right and left.

The next 40 B-25As were upgraded, with the pilot, copilot and bombardier given armor behind their seats; self-sealing fuel tanks were also installed. These reduced range, since they cut the B-25's fuel capacity from 913 US gal (3,456 lit) to 694 US gal (2,627 lit). The B-25As were the first subtype issued to an operational unit when they replaced Douglas B-18 Bolos in the 17th BG in January 1941.

The final 120 of the first production batch were delivered as B-25Bs, which brought about the first substantial armament upgrade. While the bulbous tail position was deleted, the three waist guns were replaced by two power-operated turrets mounting two .50 cal machine guns each. The upper turret, originally designed and built by Martin for its bomber, was operated by a gunner positioned within the clear plexiglass dome, with the turret mounted just aft of the trailing edge of the wing. The lower turret, directly below the upper, was unmanned, being aimed by a gunner within the fuselage using a periscopic sight and could be retracted. Unfortunately, the turret was liable to jam in the down position, while the periscopic sight was impossible to use if it got dirty.

First Blood

The B-25B was the first version of the Mitchell to serve in combat, when 16 B-25Bs, primarily from the 17th BG, participated in the famed "Doolittle Raid," but the subtype was employed primarily as an operational trainer by the United States Army Air Forces (USAAF) and Royal Air Force (RAF), the latter of which received 25 B-25Bs via Lend-Lease in August 1941, designated Mitchell I and assigned serials FK161–FK183, which were assigned to 111 Operational Training Unit (OTU) in the Bahamas. The final Mitchell from the first production batch was delivered in January 1942.

When President Roosevelt announced a production goal of 50,000 aircraft in June 1940, no one would have believed that four years later, one company alone would have produced 30,000. The massive production of war materiel in the United States was key to Allied victory. (NARA)

Planning for aircraft production, in what was seen as inevitable American participation in the war in Europe, commenced when President Franklin D. Roosevelt called for increased production of 50,000 aircraft a year to make the United States "the arsenal of democracy," following the defeat of France in June 1940. By 1941, with the USAAF and Navy having chosen the major designs for aircraft that would participate in the war, manufacturers began planning for expansion to meet the President's call for production. Such rapid industrial expansion had never been attempted anywhere. A National Defense Advisory Committee, composed of industrial executives experienced in mass production, was formed to guide the expansion. Among the members was William Knudsen, a visionary auto industry executive whose assignment was to apply automobile assembly line techniques to aircraft production. It was decided that new factories should be built in the Midwest, and several cities were selected so housing, transportation, and other urban infrastructure would not be overwhelmed. These factories were constructed of steel and concrete without windows, to enable work to continue in blackout conditions. Air conditioning was required throughout each, so a constant temperature would minimize expansion and contraction of metal, thus ensuring a better fit.

During 1940, NAA expanded its production facility in Inglewood, California, to fulfill increased orders for the T-6 trainer, while the first B-25s were produced that summer, and the first Mustang fighters the company had designed for the RAF rolled out that autumn. Working with the National Defense Advisory Committee, government-built production facilities managed by NAA were in Columbus, Ohio; Tulsa, Oklahoma; Dallas, Texas, and Kansas City (actually in Fairfax, Kansas, a suburb). The construction of the Kansas plant adjacent to Fairfax Municipal Airport in Kansas City was

One of the first nine B-25s to leave the production line in summer 1941, seen on the NAA flight line at Inglewood. The constant dihedral of the wings made these first Mitchells unstable in flight. (NARA)

approved on December 16, 1940, with formal ground-breaking on March 8, 1941. The first employees moved into their offices on April 17, and the factory was declared ready in the autumn. Eventually, 26,000 workers – the vast majority of whom had no previous experience in aircraft production – would turn out 6,883 Mitchells by the time production ceased in August 1945.

The first Mitchell subtype to be produced under wartime conditions was the B-25C/D. The B-25C-NA was built at Inglewood, while the B-25D-NC was produced in Kansas City. Initially, changes from the earlier B-25B were relatively minor. Engines were changed to the Wright Cyclone R-2600-13, which had a different carburetor, while the exhaust system went from a single pipe, exiting just behind the engine on the outer side of the nacelle, to individual stacks for the separate cylinders. The aircraft also received a cabin heater, improved instruments, and a high-pressure brake system. Due to the relatively minor changes from the B-25B, the B-25C entered production at Inglewood in January 1942, just two months after the prototype's first flight. A total of 1,625 B-25Cs were produced at the plant between January 1942 and May 1943.

The B-25C, produced at the NAA factory in Inglewood, California, was the first mass-produced B-25 subtype. (NARA)

The B-25D was essentially the same as the B-25C, with the letter designation denoting it was produced at the government-owned NAA factory in Kansas City. Construction of this facility started in late 1941, with the first aircraft leaving the production line in late 1942. (NARA)

The first B-25D-NC bomber, christened *Miss Greater Kansas City*, was completed on December 23, 1941, two weeks after the Japanese attack on Pearl Harbor, and was first flown on January 3, 1942, by NAA's Chief Test Pilot Paul Balfour. Contract NA-87 authorized construction of 1,200 B-25D-NC bombers. The first 100 were constructed from B-25C-NA parts. Following cancellation of plans to also produce the B-29 in the factory, Contract NA-100 authorized an additional 1,090 B-25D-NCs, with the first accepted on January 17, 1942. B-25D-NC production at Kansas City totaled 2,290 by March 9, 1944, when the B-25J began to roll off the line.

Several changes were introduced to the B-25C/D during the production run. First, the fuel capacity was increased to 974 US gal (3,686 lit). Soon after, the navigator received an astrodome instead of the flat window just aft of the cockpit canopy of earlier versions. About halfway through the production run, the lower turret was deleted (since it proved so problematic) and the upper turret was provided additional ammunition. Hardpoints for underwing bomb racks were added, with the outer wing strengthened to carry the extra weight. A 2,000lb (907kg) torpedo mount was fitted to enable carriage under the bomb bay, though there are no records of USAAF B-25s using torpedoes in combat (although these weapons were employed by US Navy/Marines PBJ-1s). The flexible gun in the nose was increased from .30 to .50 cal and a second .50 cal was added, firing forward from the right lower side of the bombardier's compartment and was controlled by the pilot. A 325 US gal (1,230 lit) fuel tank could be carried in the bomb bay.

Further modifications were undertaken at centers in the United States or in the field. Additional waist guns were added, initially in the rear fuselage just ahead of the tail, then later just behind the wings with larger windows that increased the field of fire, and finally with the bulged and staggered waist positions more familiar to B-25H/J Mitchells. The clear tail-cone was removed to allow a single .50 cal, and many B-25C/D Mitchells had a field-designed tail position similar to that seen in the B-25H/J, though the shallow rear fuselage prevented mounting more than one gun in the tail.

An early production B-25A Mitchell of the 34th BS "Thunderbirds" in flight in late 1941. (NARA)

Following the successful introduction of the B-25G, NAA produced the B-25H as the last version of the Mitchell built at Inglewood. It was the most heavily armed B-25 strafer, with a 75mm cannon and eight forward-firing .50 cal guns. (NARA)

Angry Noses

As early as autumn 1942, B-25s were modified in the Southwest Pacific to turn the bomber into a low-level strafer. After the utility of these conversions was demonstrated and reported by NAA field representatives involved in the effort, the company began design work on the B-25G, which would be armed with a 75mm cannon mounted in what had been the bombardier's crawl-way to the nose, with two .50 cal guns in a factory-designed "solid" nose; two additional .50 cal guns in package mounts were positioned below the cockpit on either side of the fuselage in the field. Under production contract NA-96, 400 new-build B-25Gs, with an additional 68 modified from B-25C airframes, were produced at Inglewood in May–August 1943. After completing training as a bombardier, Sterling Ditchey found himself assigned to become crew on the B-25G: "We didn't have a copilot, and the crew didn't require a bombardier, so I had to take a crash course in dead-reckoning navigation. I was the Limited Copilot, Navigator and Cannoneer." Tom Cahill described the new job in a letter:

> Whether we ride astride the cannon or sidesaddle, nobody seems to know, but already we are trying to find our new name for our trade. "Cannoneer" had a brief spurt of popularity. "Bazooka Boys" is coming up fast. "Cannigator" is a dark horse, but most popular of all is one I cannot repeat to you, my mother. We are designing a new pair of wings with crossed 75mm shells rampant on a field of dividers and protractors.

Sterling Ditchey remembered:

> The cannon had tremendous recoil normally, and to lessen that they had modified the breech so that some of the gases escaped when it fired. What that meant was it shot flame out the back. As the cannoneer, it was my job to load each round. I slammed that round into the breech and closed it, and then jumped back up on the flight deck as fast as I could to avoid that fire coming out. Even with the modification, when we fired it, it felt like the airplane slammed into a brick wall from the force of the recoil.

After successful field modification of B-25C/D Mitchells as low-level strafers in the South Pacific, NAA adapted the Inglewood-built B-25C as the B-25G, armed with a nose-mounted 75mm cannon. (NARA)

Gunner George Underwood, who flew in B-25Gs with the 310th BG in North Africa and Italy, recalled he preferred the B-25G to the later B-25H because: "The turret in the G was back in the rear. In the H it was right up behind the cockpit above the cannon and you had to get out of the way when they fired it because the backfire out of the breech could catch the gunner's flight suit on fire." Ditchey remembered they never hit the target during practice: "They had mock-up ships at Myrtle Beach south of Charleston, and we could come in low from out at sea and get off a shot or two each run, but I don't remember anyone ever hitting anything."

The B-25H immediately followed the B-25G in production, with a contract for 1,000 airframes. B-25C-10NA 42-32372 was modified as the prototype B-25H and first flew on May 15, 1943. It was powered by improved Wright R-2600-20 engines equipped with Bendix-Stromberg PR48A1 carburetors and American Bosch SF14LU-10 magnetos, with two-speed centrifugal superchargers providing a low blower ratio of 7.06 to 1 and a high ratio of 10.06 to 1 to give increased power at lower altitudes. The B-25H had the upper turret moved forward to the navigator's old position, and waist positions with wide-angle windows were placed just aft of the wing in the radio compartment; the .50 cal weapons were manned by the radio operator. The rear fuselage was deepened to allow an armored Bell M-7 tail turret with two .50 cal guns and an increased field of fire. Four .50 cal were installed in the nose above a lighter-weight 75mm T13E1 cannon fitted in the nose tunnel, while two .50 cal in factory-designed housings were mounted below the cockpit on either side. The B-25H could thus bring ten machine guns and the cannon to bear in a strafing attack.

The first B-25Hs arrived in the Pacific in February 1944, assigned to the 498th BS, 345th BG. Operational experience soon confirmed the cannon-armed B-25H offered no real advantage over strafers armed exclusively with machine guns, and use of the heavy cannon was abandoned in the Southwest Pacific by August 1944. The 498th BS passed its B-25Hs to the 38th BG (which operated B-25Gs) in September 1944; unlike most other groups, the 38th welcomed the H-model. The type was operated in the Mediterranean by the 310th BG, which flew them in concert with B-25Gs in anti-shipping operations in the Aegean and Tyrhennian seas around Sicily and Italy, from September 1943 to February 1944. The 11th BS received late-production B-25Hs with the new APG-13A radar ranging equipment for the cannon, which enabled the precise range to be determined during an attack run, making accurate aiming much simpler. The 1st Air Commando Group also flew the B-25H, most prominently *Barbie III*, flown by former American Volunteer Group ace R. T. Smith, the Deputy Group Commander.

The last B-25H was accepted in July 1944, and it was followed into production in August 1943 by the B-25J (NA-108), the version built in the largest numbers, with 4,318 manufactured in Kansas City, since Inglewood switched to manufacture of the P-51 Mustang following delivery of the last B-25H; other than the nose, the two subtypes were identical. The first B-25J, 43-3780, flew in October 1943. Between December 1943 and March 1944, Kansas City built both the B-25D and J simultaneously. In bomber configuration, the B-25J had a six-man crew: pilot, copilot, navigator/bombardier, turret gunner/engineer, radio operator/gunner and tail gunner. The bombardier nose could be replaced by a metal unit housing eight .50 cal guns. Thus modified, the designation was B-25J-11, -17, -22, -27, -32 or -37, depending in which production block the modification was made. With 18 guns, the B-25J was the most heavily armed Allied attack aircraft of the war. Frequently, the package guns on the fuselage were deleted, since they could crack the fuselage when fired.

The B-25J was the most numerous subtype built. Airframe 43-3892 was the 23rd J-model constructed, and it shows the version's combination of B-25C/D bomber nose and B-25H forward-mounted dorsal turret. Additional .50 cal guns were fitted in pairs on the fuselage sides, accompanied by extra armor plating below the cockpit windows. (NARA)

SPECIFICATIONS

The B-25J was the subtype produced in larger numbers than other subtypes with 4,390 produced out of a grand total of 9,890.

NORTH AMERICAN B-25J MITCHELL

Dimensions

Wingspan	67ft 6in (22.8m)
Length	54ft 1in (16.4m)
Height	16ft 4in (4.9mm)
Wing Area	610 sq ft (66.6m²)

Performance

Maximum speed	275 mph at 15,000ft (4,572m), 230 mph (370km/h) cruising speed.
Initial climb rate	1,110ft (338m) per minute.
Service ceiling	24,000ft (7,315m)
Range	1,275 miles (2,051km) with 3,200lb (1,451kg) of bombs. Ferry range 2,700 miles (4,345km).
Weight	21,100lb (9,570kg) empty, 33,000lb (14,968kg) normal loaded, 35,000lb (15,875kg) gross, 41,800lb (18,960kg) maximum overload.
Fuel	Four tanks in inner wing panels, total capacity of 558 US gal (2,112 lit), 429 US gal (1,623 lit) bomb bay tank. Later versions had additional auxiliary fuel tanks in outer wing panels, 105 US gal (397 lit) tanks fitted inside waist positions, 179 US gal (677 lit) self-sealing fuel tank in bomb bay, and a droppable 279 US gal (1,056 lit) metal bomb-bay fuel tank in the waist position.

Armament

Medium bomber version	Medium bomber version: One flexible .50 cal machine gun in nose, 300 rounds. One fixed .50 cal in nose, 300 rounds. Beginning with B-25J-20, a second fixed .50 cal was added in the nose.
Strafer version	Eight .50 cal machine guns in the nose with 400 rounds per gun (RPG).
All versions	Two .50 cal machine guns in individual blisters on each side of the forward fuselage, 400 RPG. Two .50 cal in top turret, 400 RPG. Two .50 cal in waist position, 200 RPG. Two .50 cal in tail turret, 600 RPG. Normal bomb load was 3,000lb (1,360kg), but a maximum bombload of 4,000lb (1,814kg) could be carried on short-range missions. Later subtypes had underwing racks for eight 5in (127mm) high-velocity aircraft rockets.

Chapter 2

Tokyo Raid
Roosevelt's Reply

The B-25 Mitchell's most famous mission was the Doolittle Raid, also known as the "Tokyo Raid," which was carried out on April 18, 1942. It was a shot in the arm to sagging home-front morale after months of defeat in the Pacific. Today, more than 75 years after the event, people not overly knowledgeable about World War Two know of the Doolittle Raid, and that the B-25 was the "star" of the show.

American military leaders began planning for a retaliatory raid against Japan within a month of the attack on Pearl Harbor, following President Franklin D. Roosevelt's statement to the Joint Chiefs of Staff, in a meeting at the White House on December 21, 1941. He believed Japan should be bombed as soon as possible to boost public morale following the Japanese assault, the so-called "Day of Infamy." The concept of launching US Army bombers from a US Navy aircraft carrier came from Navy Captain Francis Low, Assistant Chief of Staff for antisubmarine warfare, who submitted a memorandum to Chief of Naval Operations, Admiral Ernest J. King, on January 10, 1942. Low, a member of Admiral King's staff, had taken the idea to Capt Donald B. Duncan, (King's Air Operations Officer), who concluded that such an operation was technically feasible. King then passed the Low memo to Gen Henry H. "Hap" Arnold, Commander of the USAAF. Arnold called in his new special projects

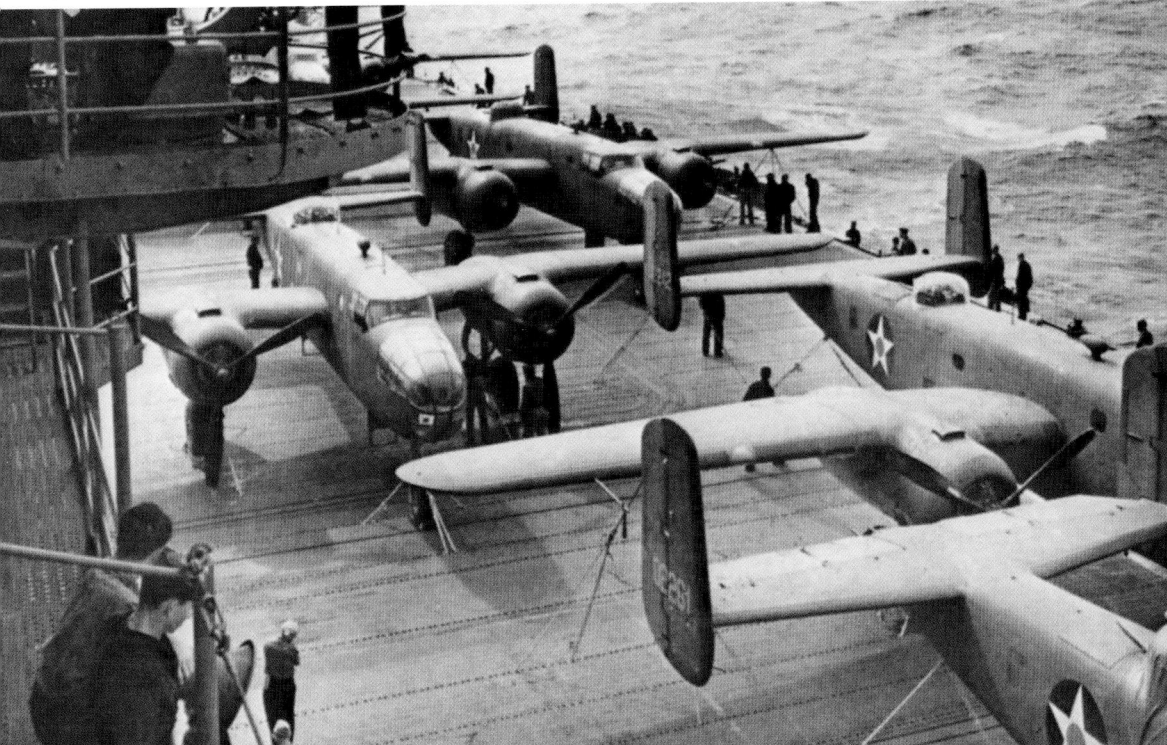

B-25B 40-2261 nearest to the camera was Ted Lawson's aircraft, while 40-2242 on the right of the photo was "Ski" York's aircraft – the latter machine was the Mitchell that made it all the way to Soviet territory and was interned. (Malcolm V. Lowe collection)

A Doolittle B-25B under guard during modification work at Mid-Continent Airlines in Minnesota, March 1942. (NARA)

officer, Lt Col James H. "Jimmy" Doolittle, America's most famous test and race pilot, and ordered him to investigate the feasibility of such a proposed attack.

Two-Force Planning

A "Tokyo project" was formed, with Doolittle and Duncan assigned responsibilities for their respective services. After reviewing the available USAAF bombers, Doolittle concluded the only type that could carry out the mission was the B-25 Mitchell; primarily due to its weight and wingspan, which would allow it to take off from the deck of a Yorktown-class aircraft carrier. The concept was demonstrated when Duncan and Doolittle successfully flew two B-25Bs off the deck of USS *Hornet* (CV-8) on February 3, 1942, confirming the basic concept's feasibility. While the bombers could take off, however, it would be impossible for them to land back on board after the mission. Doolittle's first report expressed hope they might land in Vladivostok, just across the Sea of Japan, with the bombers given to the USSR (United Socialist Soviet Republic) via Lend-Lease. This proved impossible, though, since the Soviet Union had signed a neutrality pact with Japan in April 1941 and could not afford a two-front war. Plans were then made for the B-25s to land at airfields in eastern China following the raid. From there, they would continue to western China and enter service in the Chinese Air Force.

The 17th BG (Medium), composed of the 34th, 37th and 95th BS and 89th Reconnaissance Squadron, was the first USAAF unit to take delivery of the B-25 in September 1941, and the crews were, by then, the most experienced Mitchell personnel in the Air Force. Lt Col Robert D. Knapp, who was responsible for all B-25 crew training and had personally trained the fliers of the 17th, told the men of a special "extremely hazardous" mission in which the rewards reflected the dangers. The only hint of where they were going came when they landed at Eglin Field, Florida, to meet US Navy Lt Henry L. Miller, a naval aviator who instructed them over the next few weeks in how to get a loaded B-25 airborne in just 450ft (137m), lifting off only slightly faster than stalling speed. The presence of Doolittle, renowned for his prewar exploits, impressed on them the importance of the mission. The pilots concluded they were going to hit a target in the Pacific, though most incorrectly guessed the Philippines; Doolittle told them to venture no estimates and keep their mouths shut, lest any spies learn more than they should.

Jimmy Doolittle (Left) with Captain Marc A. Mitscher (right) and seen with other raiders aboard USS *Hornet*. (NARA)

B-25B Mitchells seen aboard USS *Hornet* before their history-making mission. Note the tie-downs securing the aircraft and the cramped flight deck. (NARA)

The B-25Bs were modified extensively for the mission. Since the raid would be flown at low level, the retractable ventral turret was removed, saving 600lb (272kg) of weight, which enabled more fuel to be carried. There was no need for the still-secret Norden bombsight on such a mission, and it was removed and replaced by a makeshift version that proved more than satisfactory for low-level operations. The bomb load would be four 500lb (227kg) bombs. To deter enemy fighters attacking the rear, a pair of black-painted wooden broomsticks were attached to point through the Plexiglass tail-cone. Thus modified, take-off weight was approximately 31,000lb (14,061kg), the maximum allowable for the aircraft. Additionally, the carburetors were modified for lean operation to extend range.

Unintentional Meddling

On March 25, 22 B-25s of the original 25 flew from Eglin to McClellan Field in Sacramento, California, flying across the US at 500ft (152m) the entire distance to practice low-level navigation. Problems arose at McClellan, where the crews discovered, as they prepared to leave, that mechanics on the base, unaware of the mission, had reset the carburetors. There was no time to fix the "damage," and the raiders departed McClellan for Alameda Naval Base near San Francisco, where the men first saw USS *Hornet* tied to the dock. Just 16 of the bombers could be loaded on the vessel, though all crew members trained for the mission went aboard the carrier as back-ups. The task force steamed under the Golden Gate Bridge and headed west on April 2, 1942.

Eleven days later, north of Hawaii, Task Force 16 – commanded by Vice Admiral William "Bull" Halsey aboard USS *Enterprise* (CV-6) – rendezvoused with *Hornet*'s Task Force 18 and provided escort for the mission. Doolittle called his men to *Hornet*'s wardroom and told them: "Our target is Tokyo."

The original plan called for the bombers to be launched 400nm (740km) from Japan at dusk, so they would attack the targets at night and land in China in daylight. However, at 0738hrs on

Hornet's Captain Marc A. Mitscher watches Jimmy Doolittle's B-25B Mitchell (40-2344) take off on April 18, 1942. (NARA)

Several Japanese Medals of Friendship and Peace awarded to Americans, before World War Two, were attached to bombs dropped during the raid; following Pearl Harbor, Secretary of the Navy Frank Knox decided these should be returned! Here, Jimmy Doolittle attaches an award he received to a 500lb bomb. (NARA)

April 18, while the task force was 650nm (1,200km) from Japan, a Japanese picket boat was spotted, which radioed a warning. Doolittle and *Hornet* captain Marc Mitscher decided to launch immediately, ten hours early and 170nm (310km) farther from Japan than planned. Members of the extra crews attempted to buy their way aboard the attacking aircraft, but unsuccessfully. Doolittle's B-25, one of four carrying incendiaries, had 467ft (142m) for take-off. By 0919hrs, the 16 raiders were bound for Japan.

The crews arrived over the islands around noon, and targets in Kobe, Yokohama, Nagoya, and Tokyo were bombed, but no B-25s were lost over Japan. However, bad weather prevented the fliers from finding their landing fields in China that night, since the Chinese had not received word of their arrival; with bad weather in eastern China, they maintained radio silence, expecting a Japanese attack. Eleven crews baled out over the dark countryside, while four others crash-landed, with seven crewmen injured and three killed. The crew of one B-25B (40-2242) flew to Vladivostok due to a very low fuel state; the Soviets interned both the aircraft and crew.

Eight Doolittle raiders were captured by the Japanese and tortured to reveal where they had come from, and only four would eventually survive their capture and imprisonment. The Japanese response to the raiders having been sheltered and assisted by the Chinese was an orgy of death and destruction, as the Japanese Army initiated Operation *Sei-go* to punish the Chinese for aiding the aircrews. Chinese found with American equipment were shot immediately. Villagers were beheaded with swords, and germ warfare and atrocities resulted in as many as 250,000 Chinese being killed in the following three months.

Misjudged Result

Doolittle believed at first the mission had been a complete failure, and he told his crew he expected to be court-martialed on his return to the US; however, he was promoted to Brigadier General, awarded the Medal of Honor and assigned a new command with greater responsibility.

While all aircraft were lost and the damage inflicted was minimal, the Tokyo Raid gave an incalculable boost to American morale when nearly everything in the Pacific was going badly. The Japanese now knew their country was vulnerable to attack and kept four first-line fighter groups in the Home Islands during the fighting in the South Pacific over the next two years.

The "Doolittle Raid" was the first good news Americans received about the war after Pearl Harbor. (NARA)

More importantly, the Doolittle Raid led Admiral Yamamoto, who was busy planning the next Japanese offensive in the Pacific, to divert his attention from a possible invasion of Australia and decide on the invasion of Midway to secure the approaches to Japan. Seven weeks after the raid, the Japanese and American fleets met at the Battle of Midway, in which the core of Japanese carrier aviation was lost when four aircraft carriers were sunk. It was the turning point of the Pacific War.

CREW 5

The crew of B-25B 40-2283, 95th BS, which was fifth off *Hornet*'s deck. (L to R) Pilot Capt David M. Jones; copilot Lt Ross R. Wilder; navigator Lt Eugene F. McGurl; bombardier Lt Denver V. Truelove; flight engineer/gunner Sgt Joseph W. Manske.

Jones took off safely, despite a leak in the bomb bay fuel tank, and proceeded to Tokyo. His bombs scored direct hits on a power station, oil tanks, and a large manufacturing plant. Continuing to China, he flew on instruments until he estimated he was in the vicinity of Chuhsien, where the entire crew parachuted from the aircraft without injury. (NARA)

CREW 15

Crew 15 comprised (L to R): Pilot Lt D. G. Smith; copilot Lt G. P. Williams; navigator-bombardier Lt H. A. Sessler; flight surgeon Lt T. R. White, MD; flight engineer/gunner Sgt E. J. Saylor.

Smith and his crew were assigned targets in Kobe, southwest of Tokyo. They dropped their payload on an aircraft factory and around the dockyards, before flying towards China. Smith ditched his bomber in the waters near a small island near Sangchow. All crew members exited the aircraft safely before it sank and paddled to shore in a life raft. The Japanese hunted the area for the American raiders for days, but the aircrew evaded them in a Chinese junk. En route to Chuchow, Smith learned of Ted Lawson's serious injuries, and the evaders traveled on to meet up with him so Lt White could render medical aid. (NARA)

TED LAWSON'S CREW

(L to R): Pilot Lt T. W. Lawson; copilot Lt D. Davenport; navigator Lt C. L. McClure; bombardier Lt R. S. Clever; flight engineer/gunner Sgt D. J. Thatcher.

Lawson and his fellow crew members unloaded their bombs on industrial factories in downtown Tokyo before flying to China. Lawson was forced to ditch in the water just off the coastline and was injured severely in the crash. Just Thatcher was uninjured, and McClure was hospitalized until 1943. The crew's injuries were treated by Flight Surgeon "Doc" White, who got himself included in the raid as a gunner on the 15th aircraft (40-2267 *TNT*). He had to amputate Lawson's leg in the field before either man could join the raiders at Chuhsien. Upon his return home, Lawson wrote *Thirty Seconds Over Tokyo*. Thatcher was the next-to-last raider to die in December 2016. (NARA)

B-25 SERIAL NUMBERS

B-25 Mitchells were built in two different locations, Inglewood and Kansas City, and these were officially assigned suffix-codes to the production blocks of B-25s they built. These were NA for Inglewood and NC for Kansas City. In every instance, the military serial number of each aircraft began with the fiscal year in which it was ordered (a sequence that continues to this day). Therefore, B-25J-1-NC/43-27546 was a Block-1 B-25J built at the Kansas City facility; this particular machine was supplied to the Soviet Union under the Lend-Lease program (ultimate fate unknown).

Under the latter arrangement, the United Kingdom, Soviet Union, Brazilian, Chinese and Dutch air arms all received B-25 Mitchells; due to space constraints, just those for the RAF are listed in full. Note other six-, seven- and eight-figure serials are listed occasionally for Mitchells – these are NAA's own contract numbers, and although those for B-25Bs began with 40-*****, which may cause confusion, subsequent contracts began at 82-***** and continued (albeit non-sequentially) to 108-*****.

B-25-NA	42-87113/87137
40-2165/2188	B-25D-25-NC
B-25A-NA	42-87138/87312
40-2189/2228	42-87313/87452
B-25B-NA	B-25D-30-NC
40-2229/2348	42-87453/87612
B-25C-NA	43-3280/3619
41-12434/13038	B-25D-35-NC
B-25C-1-NA	43-3620/3869
41-13039/13296	B-25G-5-NA
B-25C-5-NA	42-64802/64901
42-53332/53493	42-64902/65101****
B-25C-10-NA	B-25G-10-NA
42-32233/32382	42-65102/65201
B-25C-15-NA	B-25H-1-NA
42-32383/32532*	43-4105/4404
B-25C-20-NA	B-25H-5-NA
42-64502/64701**	43-4405/4704
B-25C-25-NA	B-25H-10-NA
42-64702/6480***	43-4705/5104
B-25D-NC	B-25J-1-NC
41-29648/29847	43-3870/4104
B-25D-1-NC	43-27473/27792
41-29848/29947	B-25J-5-NC
B-25D-5-NC	43-27793/28112
41-29948/3017	B-25J-10/11-NC
B-25D-10-NC	43-28113/28222
41-30173/30352	43-35946/36245
B-25D-15-NC	B-25J-15/17-NC
41-30353/30532	44-28711/29110
B-25D-20-NC	B-25J-20/22-NC
41-30533/30847	44-29111/29910

B-25J-25/27-NC
44-29911/30910
B-25J-30/32-NC
44-30911/31110
44-31111/31510

PBJ-1
The following is a list of serial numbers of USAAF B-25s turned over to the US Navy as PBJ-1s with Navy Bureau Numbers:
B-25D-20-NC
41-30533/30847 as PBJ-1D 35048/35072
B-25C-20-NA
42-64502/64506 as PBJ-1C 34998/35002
42-64602/64621 as PBJ-1C 35003/35022
B-25C-25-NA
42-64708/64732 as PBJ-1C 35023/35047
B-25D-25-NC
42-87157/87180 as PBJ-1D 35073/35096
42-87181/87205 as PBJ-1D 35098/35122
B-25D-30-NC
43-3320/3344 as PBJ-1D 35123/35147
43-3570/3605 as PBJ-1D 35148/35183
B-25D-35-NC
43-3651 as PBJ-1D 35184
43-3655 as PBJ-1D 35185
43-3771/3778 as PBJ-1D 35186/35193
43-3837/3843 as PBJ-1D 35196/35202
B-25G-5-NA
42-65031 as PBJ-1G 35097
B-25H-5 n/a
43-4471 as PBJ-1H 35280
43-4482 as PBJ-1H 35281
43-4492 as PBJ-1H 35282
43-4530 as PBJ-1H 88872
43-4542/4544 as PBJ-1H 35283/35285
43-4591/4593 as PBJ-1H 35286/35288
43- 4638 as PBJ-1H 88873
43-4654 as PBJ-1H 88874
43-4655 as PBJ-1H 35292
43-4656 as PBJ-1H 35259
43-4658 as PBJ-1H 35293
43-4659/4660 as PBJ-1H 35250/35251
43-4661 as PBJ-1H 88875
43-4664/4666 as PBJ-1H 35294/35296

44-86692/86891
B-25J-35/37-NC
44-86892/86897
45-8801/9242*****

43-4667 as PBJ-1H 35252
43-4669 as PBJ-1H 35253
43-4670 as PBJ-1H 35260
43-4671/4673 as PBJ-1H 35254/35256
43-4675 as PBJ-1H 35261
43-4676 as PBJ-1H 35257
43-4682/4684 as PBJ-1H 35289/35291
43-4685/4702 as PBJ-1H 35262/35279
43-4703 as PBJ-1H 88876
43-4704 as PBJ-1H 88877
B-25H-10-NA
43-4705 as PBJ-1H 89051
43-4709 as PBJ-1H 35297
43-4710 as PBJ-1H 35258
43-4711/4883 as PBJ-1H 88878/89050
43-5028/5047 as PBJ-1H 89052/89071
B-25J-1-NC
43-3985/3986 as PBJ-1J 35194/35195
43-27511/27515 as PBJ-1J 35203/35207
43-27681/27687 to PBJ-1J 35208/35214
B-25J-5-NC
43-27904/27910 as PBJ-1J 35215/35221
B-25J-15/17-NC
44-28792/28801 as PBJ-1J 35229/35238
44-29064/29073 as PBJ-1J 35239/35248
B-25J-20/22-NC
44-29276 as PBJ-1J 35249
44-29277/29285 as PBJ-1J 38980/38988
44-29290/29299 as PBJ-1J 38989/38998
44-29510/29513 as PBJ-1J 64969/64972
44-29604/29617 as PBJ-1J 38999/39012
44-29618/29623 as PBJ-1J 64943/64948
44-29788/29794 as PBJ-1J 64949/64955
44-29801/29807 as PBJ-1J 64956/64962
44-29814/29819 as PBJ-1J 64963/64968
44-29870/29884 as PBJ-1J 64973/64987
44-29897/29901 as PBJ-1J 64988/64992
B-25J-25/27-NC
44-30353/30356 as PBJ-1J 35821/35824

44-30509/30531 as PBJ-1J 35798/35820
44-30703/30710 as PBJ-1J 35830/35837
44-30716/30718 as PBJ-1J 35838/35840
44-30849/30851 as PBJ-1J 35877/35879
44-30856 as PBJ-1J 35880
B-25J-30/32-NC

44-30961/30964 as PBJ-1J 35841/35844
44-30972/30975 as PBJ-1J 35845/35848
44-30980/30991 as PBJ-1J 35849/35860
44-31089/31104 as PBJ-1J 35861/35876
44-31277/31296 as PBJ-1J 35881/35900
44-31444/31463 as PBJ-1J 35901/35920

Royal Air Force

Aircraft from the following batches were transferred to the RAF under the Lend-Lease program:

B-25B-NA
Mitchell I FK161-FK183
B-25C-NA
Mitchell II FL164-FL218
Mitchell II FL671-FL709
Mitchell II FR393/FR397
Mitchell II MA956-MA957
B-25C-1-NA
Mitchell II FR362-FR384
B-25C-10-NA
Mitchell II FR141-FR167******
B-25C-15-NA
Mitchell II FR168-FR171******
B-25C-20-NA
Mitchel II FV900-FV919
Mitchell II FR172-FR175******
B-25C-25-NA
Mitchell II FV920-FV939
Mitchell II FR176-FR179******
B-25D-15-NC
Mitchell II FV940/FV954
B-25D-20-NC
Mitchell II FR180-FR187******
Mitchell II FR192-FR198******
Mitchell II FV955-FV979
Mitchell II KL133-KL140
B-25D-25-NC
Mitchell II FR188-FR191******
Mitchell II FR199-FR207******
Mitchell II FV986-FV999
Mitchell II FW100/FW143
Mitchell II KL141-KL145*******
B-25D-30-NC
Mitchell II FW144- FW249
Mitchell II KL150-KL161*******

B-25D-35-NC
Mitchell II HD302-HD345
Mitchell II FW249-FW280
B-25G-5-NA
Mitchell II FR208-FR209
B-25J-1-NC
Mitchell III HD346/HD361
B-25J-5-NC
Mitchell III HD362-HD380
B-25J-10/11-NC
Mitchell III HD381-HD383
B-25J-15/17-NC
Mitchell III HD384-HD400
Mitchell III KJ561/KJ614
B-25J-20/22-NC
Mitchell III KJ624/KJ731
B-25J-25/27-NC
Mitchell III KJ732/KJ771
B-25J-30/32-NC
Mitchell III KJ772/KJ800********
Mitchell III KP308/KP328 – all returned to USA

Notes

* B-25C-15-NA 42-32384/32388 modified as B-25G-1
** B-25C-20-NA 42-64531, 64558, 64561, 64563, 64569, 64579/64582, 64584/64587, 64649, 64654, 64668, 64670/64675, 64692, 64693, 64696/64701 converted to B-25G
*** B-25C-25-NA 42-64702/64707, 64753/64772, 64779, 64780 converted to B-25G
**** B-25G-5-NA 42-65031 to USN as PBJ-1G 35097
***** B-25J-35-NC 45-9000/9242 never built; contract cancelled
****** Served with 320 (Dutch) Squadron
******* Retained in Canada for the Royal Canadian Air Force
******** Mitchell III (B-25J) serials KJ774, KJ777/ KJ783, KJ785, KH787/KJ792, KJ795/KJ799 diverted to USAAF

B-25 NOSE ART

Above left: B-25J-2, 43-27629 *San-Antoneli Rose* of the 486th BS/340th BG, while stationed at Alesani, Corsica. (NARA)

Above right: B-25J, 43-27636/III *Ave Maria*, Lt W. E. Marchant, 447th BS/321st BG, Solenzara, Corsica. (NARA)

Above left: B-25C, 41-12860 *Desert Warrior*, Capt R. Lower, 81st BS/12th BG, Bolling Field, Washington DC, July 1943 (War Bond tour). (NARA)

Above right: B-25C, 41-29956 *The Black Widow*, 82nd BS/12nd BG. (NARA)

Above left: B-25J-2, 43-27637 *Big Noise*, Lt R. McEldery, 428th BS/310th BG, Ghisonaccia, Corsica. (NARA)

Above right: B-25J *Hardships 2nd*, 38th BG, seen at Clark Field, Philippines, 1945. The artwork encompasses the unit badges of the 38th's component squadrons – 71st, 405th, 82nd and 823rd BS. (NARA)

Chapter 3

Export Users
Worldwide Mitchells

Although the USAAF was the biggest user of B-25 Mitchells, the type was also operated in significant numbers by two other main air forces during World War Two, the RAF and the Soviets, as well as smaller Allied groups.

The RAF ultimately received 900 examples of the medium bomber, where it was known just as the Mitchell. These included 23 B-25Bs (Mitchell Mk.I), 167 B-25Cs and 371 B-25Ds (Mitchell Mk.II) in 1942–44, and 316 B-25Js (Mitchell Mk.III) between August 1944 and August 1945. The latter were issued as replacements for 2 Group's Mitchell IIs from November 1944 (after most of that formation's assets were transferred to the 2nd Tactical Air Force), although two squadrons retained the Mitchell II, as this lighter variant was regarded as possessing better control response. These were the only B-25s operated in the United Kingdom and northwestern Europe, as the USAAF had decided upon the Martin B-26 Marauder for this theater, although Mitchells did operate in the Mediterranean (see Chapter 5).

As snow lies on the ground, 1,000lb bombs are loaded into an RAF Mitchell bomber at a Belgian airfield, on January 15, 1945. (All images in this chapter are Key Collection unless otherwise stated)

The initial batch of Mitchell Is were delivered in August 1941 and were used exclusively for training and familiarization by 111 OTU in the Bahamas, and they were never flown on operations. The first combat-capable B-25s were Mitchell IIs, and served with 2 Group, which was the RAF's tactical medium-bomber force, where they replaced Lockheed Venturas, as these were unsuited to conditions over northwestern Europe. Mitchells were assigned to four squadrons (98, 180, 226 and 329) in 139 Wing, and commenced operations in early 1943.

Six aircraft from 98 and 180 squadrons constituted the RAF's first combat mission with the type, when they attacked oil installations at Ghent (occupied Belgium) on January 22, 1943. One bomber was shot down over the target, with two more lost to Fw 190s on the return leg; as a result, both squadrons were stood down temporarily to improve tactics, before operations resumed on May 13, 1943. The result was to operate in tight six-ship formations, which had the effect of concentrating bombing efforts as well as defensive fire against enemy fighters. Unlike their USAAF counterparts, RAF Mitchells retained the retractable ventral turret, since protection from below was essential for European operations.

Supporting the Invasion

By this time, the Mitchells had participated in Exercise *Spartan*, which occurred in March 1–12, 1943, and was the largest war exercise ever held in Britain. This tested the ability of the RAF to provide close support to the army, utilizing tactics and operations developed in North Africa and Italy by the Desert Air Force, led by Air Vice Marshal (later Air Marshal) Arthur Coningham. The exercise led to the formation of the 2nd Tactical Air Force (2TAF) in June 1943 (with assets transferred from 2 Group in July) and was to provide dedicated support to the British 2nd and Canadian 1st armies during and after D-Day.

Preceding the invasion, Mitchells flew day and night operations from RAF bases at Dunsfold, Swanton Morley, Hartford Bridge, and Foulsham against marshalling yards, bridges, and roads; V-1 launch sites were added in late 1943 and V-2 sites after D-Day. Crews were formed from many nationalities, with British and Commonwealth (Australian, Canadian, South African and New Zealander) personnel joining with those who had escaped from western Europe (Czech, Dutch, Polish and French).

Following the invasion, Mitchell squadrons concentrated on isolating enemy forces from reinforcements and resupply. Aircraft from 226 Squadron Special Signals Flight, codenamed "Ginger Flight," began missions from the night of June 1, flying lone aircraft intelligence-gathering missions at 20,000ft (6,096m) deep in enemy territory, where they received radio reports from the French resistance regarding positions of German units for later attack.

Mitchell IIIs (B-25J) were equipped with a tail turret mounting two .50 cal weapons.

Panzer Group West Strike

While many spectacular daytime strikes were carried out by 2 Group during the war, the most important (yet least publicized) was arguably the "Dinner Raid" mission on June 10, 1944, which involved the use of intelligence gathered at Bletchley Park. This material, known as "Ultra," was obtained through decryption of Enigma signals traffic (which the Germans believed to be unbreakable), and was so sensitive that, to protect the source, a cover story was devised claiming the attack was based on information from the French Resistance and confirmed by aerial reconnaissance.

During the immediate post-invasion phase, Field Marshal Rommel had pleaded with Hitler to deploy the Panzer Group West reserves, commanded by Gen Geyr von Schweppenburg, which was finally approved on June 9. After Ultra-decrypted traffic had revealed this information to the Allies, high-frequency direction-finding equipment located Panzer Group West's headquarters at the Chateau de la Caine, southwest of Caen, and a maximum-effort strike was organized. Seventy-two Mitchell IIs and IIIs (each carrying eight 500lb bombs) departed RAF Dunsfold at 2022hrs on June 10, accompanied by 34 rocket-armed Typhoons from 181, 182, 245 and 247 squadrons and escorted by Spitfires from four squadrons.

Panzer Group West's staff, including Chief of Staff Gen von Dawans, were at dinner when the sirens sounded at 2115hrs; all went outside to watch and only realized they were the target when the first wave of Typhoons dived to attack. While the fighter-bombers fired 136 rockets, the Mitchells dropped 536 500lb (228kg) bombs, hitting the chateau and the surrounding area. Gen von Schweppenburg was wounded when he arrived in his staff car, just as the attack began, while von Dawans and his entire staff perished; as a result, Panzer Group West headquarters ceased to exist. Preparations for a German counterattack were delayed for three weeks while a new staff was assembled, but by then it was too late.

This Mitchell II, FL218/EV/W of the RAF's 180 Squadron, was based at Foulsham in Norfolk during 1943; the aircraft was hit by flak on January 25, 1944, and crashed near Hawkinge, Kent. Its pilot, W/O D. Rogers, was killed, but the other three crew members baled out. Its nose art title "Nulli Secundus" translates as "Second to None."

Mitchell Mk.II FW172 wearing "Invasion Stripes" takes off from RAF Dunsfold during August 1944. Coded "EV-V," the aircraft belonged to the 2nd Tactical Air Force's 180 Squadron, RAF. (John Batchelor Collection via Malcolm V. Lowe)

Advancing into Europe

Another notable, but more-costly, mission occurred on July 23, when 15 Mitchells from 98 Squadron attacked a railyard at Glos-Montfort, employing the Gee-H radio-navigation aid, due to complete cloud cover over the target. The mission was led by Wing Commander Pau, the second box formation by Squadron Leader Paynter and the third by Flight Lieutenant Brown, with "bombs gone" called at 1652hrs. Moments later, Paynter's Mitchell, *S-Sugar*, exploded, and Flt Lt Weekes' aircraft *R-Roy* caught fire – it was last observed diving into the clouds. Flying Officer Berry's *G-George* was also damaged and forced to land in the American sector with three wounded crewmen, while Fg Off Harris' *H-Harry* landed at RAF Tangmere, with one crewman injured badly. All missing airmen were determined to have been killed. An investigation determined Paynter had jettisoned his bomb load, probably due to intervalometer problems, and some of the bombs had collided and exploded directly below the aircraft. A directive was issued subsequently that problem bomb loads were only to be dropped over the English Channel, and away from the main formation.

In September 1944, the four Mitchell squadrons moved to airfields in Belgium and France, where they could operate in closer support of the 21st Army Group (the British headquarters formation overseeing the British 2nd and Canadian 1st armies). Bombing missions were flown at all hours to isolate enemy forces in western Germany, with Mitchells and Mosquitos often dropping marking flares for each other's night raids against troop and supply movements.

The final large loss of RAF Mitchells to enemy action occurred on January 1, 1945, when 13 aircraft from 80 Squadron were destroyed on the ground by the Luftwaffe during Operation *Bodenplatte* (*Baseplate*), in which 16 Allied airfields were targeted by approximately 500 German fighters/fighter-bombers.

Mitchell IIs were also operated by 320 (Dutch) Squadron, following its re-equipment in September 1943, after assignment to 2 Group. Originally, it had been formed in June 1940 with personnel from the Royal Dutch Naval Air Service and had flown Fokker T.VIIIs and Hudson IIIs on antisubmarine and anti-shipping operations over the English Channel and North Sea. As part of ongoing efforts to liberate Europe, the squadron moved to Belgium in October 1944 (by which time it had also received Mitchell IIIs) to fly tactical missions as part of 139 Wing until Victory in Europe (VE) Day.

While not as numerous as its European Theater compatriots, the RAF flew Mitchell IIs in the Far East, initially with 3 Photo-Reconnaissance (PR) Unit, which was subsumed into 681 PR Squadron, before

Above: A string of bombs is dropped by a Mitchell II; in this case, the subject is FV914/VO-A of the RAF's 98 Squadron.

Left: The Mitchell III (B-25J) was fitted with an Emerson upper turret, rather than the Martin-built unit seen on the Mitchell II (B-25C/D).

the latter's twin-engined assets were transferred to 684 PR Squadron. Photo-reconnaissance sorties were flown from Dum Dum airfield, located near Calcutta, India, from the spring of 1942 onwards.

As late as December 1945, there were 393 Mitchell IIs and IIIs still on RAF inventory, though all were subsequently disposed of during 1946.

This late-production Mitchell II (B-25D-30) could be identified by the factory-installed, manually operated single-gun tail position, and the large waist windows immediately aft of the wing on both sides, with a .50 cal weapon in each.

Mitchell IIs of 180 Squadron RAF, most likely at this unit's station at Foulsham, Norfolk, UK. These airframes conform to baseline B-25C format.

An RAF Mitchell II flies over shipping in the English Channel, in June 1944. Note the style of "invasion" stripes.

A Mitchell II over Normandy following the Allied invasion. Note the 24in (60cm) black/white ID stripes worn by multi-engine aircraft.

The Southwest Pacific

In addition to the Mitchell IIs operated by 320 (Dutch) Squadron in Europe, the Netherlands operated the Mitchell in the Southwest Pacific. On June 30, 1941, the Netherlands Purchasing Commission signed contract 71311/NA with NAA for 162 B-25C aircraft, with the first to be delivered in November 1942, and the remainder by February 1943. In September 1941, three B-25s were released to the Dutch for training, and the delivery schedule accelerated, with 42 to be delivered in March–September 1942, 36 in October–November, 72 in December, and the last 12 delivered in February 1943.

However, in the aftermath of Pearl Harbor, the Dutch asked for this process to be expedited further, and on January 21, 1942, an emergency allocation of 60 B-25s was approved for the Netherlands East Indies Air Force (NEIAF) at Archer Field, Australia, and Bangalore, India. Unfortunately, these aircraft arrived too late to help stem the Japanese advance, since the Netherlands East Indies capitulated to the Japanese on March 9.

Those NEIAF B-25s that had arrived (serials N5-139, N5-143, N5-144, N5-145, and N5-148) were requisitioned by the RAF and modified for photographic reconnaissance, all being operated by 681 PR Squadron. Two (N5-139, N5-143) were assigned RAF numbers MA956 and MA957, but the other three retained their NEIAF identities.

Similarly, Dutch B-25s delivered in Australia were soon "requisitioned" by the USAAF as the new 5th Air Force was organized. The next batch of B-25Cs were also seized by the USAAF. In August 1942, 18 Squadron, which had been formed as a Dutch-manned, Royal Australian Air Force (RAAF)-directed unit, finally received its "permanent" supply of Mitchells. While most crew were Dutch or Australian, there were 38 nationalities, speaking 13 different languages. In December 1942, the squadron moved to MacDonald in Northern Territory and began operations. Eventually, a total of 150 Mitchells was assigned to the NEIAF, 19 in 1942, 16 in 1943, 87 in 1944, and 28 in 1945.

The more capable B-25J Mitchell replaced most of the earlier models in 1944, and the squadron was well known for its anti-shipping strikes until fighting ended in the Pacific with the Japanese surrender.

On January 17, 1946, the unit transferred formally from Australian to Dutch control and became a unit of the Royal Netherlands East Indies Army Air Force (KNIL-ML); its Mitchells were soon involved in combat against the Indonesian independence movement. Both glass- and gun-nosed B-25Js were

Following the end of the Indonesian independence struggle in 1949, ex-Dutch Air Force B-25Js were transferred to the Indonesian Air Force, where they were operated in secondary roles until the mid-1960s.

operated by 18 Squadron, which was joined by the newly formed 16 Squadron in November 1946 with nine B-25Js, the latter operating from Palembang until it was merged with the former in August 1948.

Twelve B-25s were operated by a conversion unit at Piak from mid-1946 until August 1948, retraining former POWs and new pilots from The Netherlands. A photographic reconnaissance unit was formed in January 1947 with a further five B-25s; this too was merged with 18 Squadron in August 1948. Fighting in Indonesia continued until the armistice was agreed on in July 1949. The Republic of Indonesia was declared on December 27, 1949, and 18 Squadron was disbanded in June 1950, with its B-25s transferred to the new independent Indonesian air force, the Angkatan Udara Republik Indonesia (AURI). After its disbandment in August 1945, 320 Squadron was reformed in 1949 as a Royal Dutch Navy maritime patrol squadron and was equipped initially with B-25s, among which were some of the very same aircraft it had used in World War Two.

Mitchells Down-Under

The RAAF was late in adopting the Mitchell, and it operated relatively few examples of the type. By the spring of 1944, 18 Squadron had more than enough airframes to spare a few. Arrangements were made to re-equip 2 Squadron RAAF, with the first batch of 20 B-25Ds (serials A47-1 to A47-20) transferred to the RAAF in late April 1944. In June, five B-25Ds (A47-21 to A47-25) and seven B-25Js (A47-26 to A47-32) were transferred, while a further five B-25Ds (A47-33 to A47-37) and two more B-25Js (A47-38 and -39) were delivered in August and September, respectively. The final 11 B-25J (serials A47-40 to A47-50) were assigned directly from the USAAF to the RAAF between April and August 1945, but just two (A47-41 and A47-43) were allotted to 2 Squadron.

The Mitchells replaced the unit's Beauforts, and the squadron flew sorties with the Dutch-operated Mitchells of 18 Squadron against targets in the East Indies, where both formations achieved considerable success in the anti-shipping role. Operations continued until November 14, 1945 (although 2 Squadron was not disbanded until May 1946), with the Mitchells placed into storage until March 1950, when most were subsequently scrapped.

Canada was a postwar operator of the B-25 series; this example, with overpainted nose glazing, is a Mitchell II of 418 (City of Edmonton) Squadron.

Canadian Operations

While many Canadian aircrew flew Mitchells during World War Two, the Royal Canadian Air Force's (RCAF) association with the type was during the postwar years, and its first B-25s were originally diverted from RAF orders. These included one Mitchell I, 42 Mitchell IIs, and 19 Mitchell IIIs. Several Mitchell IIs were operated by 13 (P) Squadron RCAF, which was formed unofficially at Rockliffe in May 1944 to undertake the high-altitude aerial photography role. The unit was redesignated 413 (P) Squadron in April 1947 and continued to fly Mitchells until October 1948.

Further Mitchell IIs were allocated to 418 (Auxiliary) Squadron in January 1947 and later also some Mitchell IIIs, which were used until March 1958. Other squadrons operating the twin-engined bomber included 406 (Auxiliary) Squadron with Mitchell IIIs from April 1947 to June 1958, and Air Transport Command's 12 Squadron, which also flew Mitchell IIIs, but from September 1956 to November 1960. The RCAF received an additional 75 B-25Js from USAF stocks in 1951 to make good attrition losses and equip other second-line units.

From Asia to South America

China received 100 B-25C/D Mitchells, along with 131 B-25Js under Lend-Lease. Four squadrons of the 1st Medium Bomber Group, which had previously operated Russian-built Tupolev SB-2 bombers, were reequipped with B-25s beginning in 1943. Notably, the 1st Bomb Group was part of the Chinese-American Composite Wing (Provisional), a joint USAAF and Republic of China Air Force organization, subordinate to the 14th Air Force during World War Two. Postwar, the four bomber squadrons retained their Mitchells, which were used in the Civil War until the Communist victory in October 1949.

Mitchells were also operated by a Free French Air Force-manned unit, 342 Squadron RAF, originally equipped with the Boston light-attack bombers, but reissued with 21 Mitchell IIIs in 1945. Postwar, the unit was transferred to the Armée de l'Air as Groupe de Bombardement I/20 "Lorraine." While the Mitchells were retained in service with the French after the war, with a number converted to VIP transports, they were finally struck off charge in June 1947.

Brazil was the only Latin American country to operate the B-25 during World War Two, with the Força Aérea Brasileira receiving seven B-25Bs, one B-25C, 11 B-25J-15s, and ten B-25J-20s (all B-25Js were delivered between August and November 1944). These Mitchells were operated by the No 4 Grupo de Bombardeio Médio, based at Foraleza, while the Brazilian Navy's No 6 Regimento de Aviação's 2 Group operated a mix of B-25s, PV-1s and A-28s on antisubmarine patrols. After the war, 64 additional B-25Js were delivered between July 11, 1946, and October 2, 1947, under the American Republic Projects and subsequent military assistance programs. Several operational squadrons, base flights, and air depots flew the B-25 as transports until its final retirement in 1974.

On the Eastern Front

The Soviet Union was the third-largest B-25 operator, with 862 B-25s (B, D, G, and J subtypes) supplied under Lend-Lease during World War Two. A Soviet Air Force delegation, led by test pilot M. M. Gromov, arrived in the US on August 5, 1941, to study and select bombers to be delivered to the USSR for evaluation, regarding further supply under Lend-Lease. The B-17 was the team's initial preference, but the US refused to supply these. After visiting NAA and the Glenn L. Martin Company, the delegation agreed on three B-26 Marauders and two B-25B Mitchells on September 16, 1941. After further study and review, the final decision at the end of the month was for five B-25Bs, which were considered superior to the early B-26, if additional deicing equipment could be provided. Subsequently, the first two B-25Bs were delivered to Murmansk by ship on December 20, 1941, as part of convoy PQ-6, and both aircraft arrived in Moscow at Monino Airfield on March 5, 1942.

Soviet Lend-Lease aircraft were delivered originally via what was known as the "Southern Route," with aircraft flown from the US to Brazil, across the South Atlantic to Liberia, then across Africa to Cairo, and on to Abadan, Iran, where they were met by Soviet crews and flown into the USSR. An alternative transit path, the Alaska–Siberia ferry route, opened in late 1942, and from 1943 onwards, all Mitchells were transferred via this way.

Soviet crews were impressed with the modernity of the aircraft initially, and they remarked favorably on the quality of the comprehensive operation and maintenance manuals, even though they were in English. Notably, the use of cartoons, which warned of problems to be avoided and were understandable regardless of the language spoken, were much appreciated by the Soviets.

Unsurprisingly, the B-25s required considerable modification to enable their use in the Russian winter, including altering the R-2600 engines to use oil thinned with aviation gasoline, to facilitate cold starts. Additionally, there were failures of individual components, including electrical instruments and armament, frozen and burst hydraulic brake lines, frozen controls for raising/lowering the landing gear, flaps, and opening/closing the bomb bay doors. Oil coolers cracked, tired engines misfired due to spark plug failures, cracks appeared in plug leads, and carburetors failed.

The cold created serious problems with the fuel and oil cells, which were composed of several layers of rubber with varying physical characteristics; these reacted differently to the weather. Inner liners would crack, and pieces of dissolved raw rubber would flow into the fuel cells and, more importantly, plugged the holes in fuel and oil lines, and the fuel filters. The replacement of these rubber cells was complicated, time-consuming, and back-breaking work, which involved partial disassembly of the wing skin to gain access. Throughout the war, the Soviets found it necessary to produce replacement parts for the B-25s in the USSR, due to difficulties in obtaining timely supplies from the US.

The first Mitchells were allocated to the 37th Bomber Aviation Regiment (BAP) of the 222nd Bomber Aviation Division (BAD), soon followed by two other regiments, the 16th and 125th BAP. Each regiment comprised three squadrons of 12 aircraft each. Initially, the division was assigned to the 1st Bomber Aviation Army (BAA) of the Voyenno-Vozdushnye Sily (Military Air Forces, VVS),

The Soviets solved the problem of warming B-25 engines in the depth of a Russian winter via factory-supplied heating covers, and gas-powered generators.

where the aircraft were flown on low-level battlefield interdiction missions in daylight. However, they suffered such severe losses that it was feared replacement aircraft (flown from the US), would be insufficient to cover such attrition.

In July 1942, the 222nd BAD became part of the Reserve of the Supreme High Command. Following the high losses, division leaders demonstrated that, if the Mitchell was flown at 10,000–12,000ft (3,050–3,660m), it had a range of 930 miles (1,500km), which was considered long distance by the Soviets. On September 29, 1942, the 222nd BAD was again transferred, this time to Long-Range Aviation (ADD), where it flew night missions against the enemy's rear echelons, equipped with additional fuel tanks.

By late 1942, the supply of Mitchells enabled the units to be expanded, with the 222nd BAD becoming the 4th Guards Aviation Division (GAD), with more B-25s assigned to its component bomber regiments. On July 3, 1943, just before the Battle of Kursk, the 4th GAD was supplemented by two new Mitchell-equipped regiments in the 5th GAD, with the two divisions now part of the 4th Guards Aviation Corps (GAK). This formation was the main Soviet Mitchell operator in the ADD; a third regiment was added to each division during the winter of 1943/44.

Onto the Offensive

The Soviets particularly liked the B-25D-25 and D-30 Mitchells, which began arriving shortly after the Battle of Stalingrad in 1943. Even at this stage of the war, changes were made to the airframes, notably the removal of the remote lower turrets, which had proven impossible to maintain. The D-25/30 Mitchells had large waist windows immediately aft of the wing, with a .50 cal machine gun in each, and a factory-supplied tail gunner position with a single machine gun. These bombers were considered the most "survivable," until the arrival of the better-armed B-25Js in 1944 (note, by January 1944, Mitchells constituted one-tenth of the ADD's total bomber force).

In addition to supporting almost all major operations of the Red Army in 1943–1945, the 4th GAK played a significant role in the strategic bombing attacks against Helsinki in February 1944. In September 1944, the 5th GAD supported the liberation of Belgrade; five pilots from this division were decorated as Heroes of Yugoslavia. Also that month, ADD units supported the ultimately unsuccessful Slovakian National Uprising (which had begun on August 29) with an air bridge comprising Li-2s, C-47s and B-25s. The 4th GAK aircraft dropped supplies to Czech air units at Tri Duby airfield in central Slovakia, commencing on the night of September 4/5, 1944. In total, 498 sorties transported 253 tons of arms and ammunition to the insurgents for the loss of one aircraft, before the evacuation of Tri Duby at the end of October. Mitchells also flew support missions for Soviet partisan forces in Eastern Europe after the Battle of Kursk to the end of the war.

After major restructuring in January 1945, the ADD was transferred to the VVS and redesignated as the 18th Air Army (VA), which by that time had 320 Mitchells, constituting one-fifth of its total strength. By June 1946, just 252 B-25s remained on the inventory as the 18th VA was reorganized into the Long-Range Aviation of Air Forces (Dal'naya Aviatsiya Vozdushnikh Syl, DA VS).

At the end of the war, several B-25J Mitchells equipped with S-1 autopilot systems and Norden M-9 bombsights were delivered; the top-secret and restricted Norden system had long been at the top of Moscow's "wish list." As the Soviets advanced into Germany, many German Lotfe-7 bombsights were captured, and studies were conducted in late 1945 to fit these on B-25s, but these tests resulted in directional instability on the modified aircraft (due to incompatibility with the bomber's autopilot).

Above: A Soviet B-25D, of the 13th GAP, pictured during the winter of 1943/44 at Novo-Dugino. Obvious details include canvas covers to protect the engine oil from freezing and roughly applied whitewash camouflage. (Aleksandr Dudakov Archive)

Right: The refueling of a Soviet Air Force B-25D takes place at Monino Airfield. (Aleksandr Dudakov Archive)

Postwar in the USSR

While most Lend-Lease aircraft in Soviet service were destroyed between 1945 and 1947, under the supervision of American inspectors, Mitchells were used extensively for several years after the war and were known in the West by the NATO reporting name of *Bank*. The Bobruysk-based 330th BAP used Mitchells from 1946 until conversion to Tupolev Tu-4 *Bulls* (reverse-engineered B-29s) in 1949, while the 132nd BAP (stationed at Sakhalin) converted to from Tu-2s to B-25Js in 1950, and a Kamchatka-based regiment still operated the type in 1953.

As late as 1951, the DA VS's 121st Independent Guards Reconnaissance Aviation Regiment (GRAP), previously a Guards Bomber Aviation Regiment, converted to reconnaissance-modified Mitchells, and was subsequently redesignated the 121st Independent Guards Long-Range Reconnaissance Aviation Regiment (GDRAP). This unit operated B-25s until 1953, when new Tu-4 bombers were received.

In the early 1950s, the test pilots' school at Kratovo (now Zhukovskiy) still had several B-25s on its inventory. Due to its excellent flying characteristics and a strong structure that allowed extensive modification, the Mitchell was a popular test bed for various Soviet research and development projects in the late 1940s, including flight trials of early jet engines such as the RD-10F (a Jumo 003 copy), and the study of air-refueling methods.

Throughout its operation by the Soviets, the Mitchell was extremely popular with the crews. Aleksandyr Dudakov, who later rose to the rank of Major General of Aviation, had been an instructor pilot for three years when he was assigned to fly the Mitchell. After one familiarization flight at Monino, and a longer mission the next day, he was considered "checked out" on the type and assigned as a first pilot. Dudakov would fly the Mitchell from just before the Battle of Kursk in July 1943 to the end of the war. He later recalled:

> The B-25 was such a nice and simple airplane, that I think it was easier to fly than the U-2 [Po-2] trainer. With two engines and two tail fins, it had great control with one engine out. It was an obedient, very reliable airplane. Our best engines worked for 300 hours, while the Wright-Cyclones on the B-25 had a life up to 500 hours. I loved this airplane. It was my luck that I was sent to fly it.

The crew of a Soviet 15th GBAP (Guards) B-25D pose for posterity; the patriotic slogan 'Sevastopol'skiy' is noteworthy... a feature seen on many Soviet fighters and bombers during World War Two. (Aleksandr Dudakov Archive)

Chapter 4

Gun Noses
The Strafers

The story of the Mitchell strafer centers on gifted self-taught troubleshooting engineer and "tinkerer extraordinaire," Paul Irvin "Pappy" Gunn. His nickname was given by young GIs assigned to work for him, who thought Gunn "old" at the advanced age of 43. Gunn was a veteran enlisted United States naval pilot, who had retired in December 1939, after 21 years' service, and started Philippine Air Lines. Following the Japanese attack on Pearl Harbor, he received a wartime commission as a captain in the USAAF.

Gunn first encountered the Mitchell on April 11, 1942, when nine B-25Cs and three B-17Es, led by BG Ralph Royce, arrived at Del Monte Airfield on Mindanao. Gunn's experience was crucial for them to stage attacks over April 11–13 as they made the Mitchell's combat debut, hitting shipping and harbor facilities at Cebu, the harbor and airstrips at Davao, and Nichols Field on Luzon. When they returned to Darwin, Capt Gunn was among those they brought with them.

Six months later, Gunn (by this time a major) was in New Guinea with the 3rd Bombardment Group, which flew A-20A Havoc twin-engined light bombers. Using weapons salvaged from battle-damaged P-40s, he mounted four .50 cal machine guns in the bombardier's position, turning the A-20s into strafers. Fifth Air Force commander Maj Gen George Kenney put Gunn in charge of special projects,

The B-25H Mitchell appeared in summer 1943. With ten forward-firing .50 cal machine guns and a 75mm cannon, it was the most heavily armed Mitchell gunship. (Key Collection)

and he turned his experimental up-arming efforts to the B-25, converting the medium bomber to a gunship, armed with an extra four .50 cal machine guns in the nose, and four more in two-gun "packs" on either side of the fuselage, below the cockpit. With the bombers so modified and crews trained for low-level attack, the gunships were ready for their big test, the Battle of the Bismarck Sea on March 2–4, 1943.

Tested in Battle

Following the loss of Buna (in the Territory of Papua, now Papua New Guinea) in January 1943, the Japanese needed to reinforce nearby Lae (in Northeast New Guinean territory). Despite the threat posed by Allied air power, a large convoy sailed from Rabaul (on New Britain Island) direct to Lae with 6,900 troops. However, Japanese radio traffic was intercepted and decoded and, from the intelligence gleaned, plans were made to interdict the reinforcements. The eight destroyer escorts and as many troop transports departed Rabaul on February 28, 1943, under cover of a storm, but a patrolling B-24 Liberator spotted them the next day, and the convoy was shadowed until it came within range of aircraft from Port Moresby.

Sustained air attacks on the convoy were launched on March 2/3, involving RAAF Beauforts, Beaufighters and A-20s, along with USAAF A-20s and B-25s from the 38th BG. Copilot Garrett Middlebook described the results:

B-25C 41-12971 *Dirty Dora* **at Three Mile Aerodrome, Port Moresby, New Guinea, December 1943. In December that year, B-25 gunships of the 398th BS "Falcons" and 399th BS "Bats Outta Hell" (both 345th BG) were painted with spectacular nose art, in recognition of the units' leading roles in local ground-attack operations. (Tatelman Archive via Adam Lewis)**

B-25H Mitchells of the 1st Air Commando Group (ACG) supported the Chindits' raid into Burma in 1944. The aircraft were led by former American Volunteer Group "Flying Tiger" ace R. T. Smith. (Key Collection)

They went in and hit this troop ship. What I saw looked like little sticks, maybe a foot long or something like that, or splinters flying up off the deck of ship; they'd fly all around and twist crazily in the air and fall out in the water. Then I realized I was watching human beings. I was watching hundreds of those Japanese blown off the deck by those machine guns. They splintered around the air like sticks in a whirlwind and they'd fall in the water.

The final score was all eight transports and four destroyers sunk; not only had the convoy and its cargo been destroyed, but the concept of medium and light bombers as gunships had been validated emphatically. Due to this success, Maj Gunn was placed in charge of modifying all A-20 and B-25 aircraft in the Fifth Air Force to gunships. One of his more outlandish approaches was to equip B-25s with a stripped-down 75mm pack howitzer, which he demonstrated personally in attacks on Japanese shipping. Gunn's heavy weapon designs were sent to NAA and adapted quickly, resulting in the B-25G, while his earlier efforts led to the rapid development of the solid-nosed B-25H and B-25J. The latter's factory-designed eight-gun nose made it the most heavily armed medium bomber of the war, with no fewer than 14 forward-firing .50 cal machine guns.

Rampage Across the Southwest Pacific

The 38th BG, known as the "Sunsetters," arrived in Australia at the end of February 1942 and became the first all-strafer B-25 group by early 1943. The "Sunsetters" were followed in May 1943 by the 345th BG "Air Apaches," which comprised four squadrons: 498th BS "Falcons," 499th BS "Bats Outta Hell," 500th BS "Rough Raiders," and 501st BS "Black Panthers." The BG made up for its late arrival in the two years between its first mission on June 30, 1943, and the end of the war, by becoming the leading south Pacific B-25 attack unit.

A B-25J Mitchell gunship of the 501st BS, 345th BG "Air Apaches," on Ie Shima Island in July 1945, is armed prior to an anti-shipping mission. A sustained burst of fire from the 12 fixed forward-firing .50 cal guns could shred many a vessel. (Key Collection)

In August 1943, First Lieutenant Vic Tatelman reported to the 345th BG, assigned to the "Bats Outta Hell." Tatelman would become the only man to complete two tours with the "Air Apaches," becoming the leading B-25 attack pilot of the war. He was assigned B-25C 41-12971, a 38th BG hand-me-down veteran of the Battle of the Bismarck Sea. Named *Dirty Dora* by a previous pilot, for an Australian woman he met in Sydney, the battered old bomber would become legendary.

Attacking Rabaul

One of the 345th's most outstanding missions was the attack on Rabaul on October 18, 1943, which became the opening bell of the air campaign that eventually isolated and bypassed the base. When the Australian liaison officer said: "When you go down…ahhh, if you go down – make your way to this location." As he pointed to the map, no one laughed. The mission involved two groups of B-24s with P-38 escorts to attack 30 minutes before the Mitchells arrived, and to draw in defending fighters so the 345th and 38th BGs could catch the Zekes (Mitsubishi A6M Zero) refueling.

Thirty-six Mitchells departed Dobodura, led by Lt Col Clinton L. True in B-25D 41-30024 *Red Wrath*; Rabaul was 200 miles (322km) away, across the Bismarck Sea. Unbeknownst to the crews, the bad weather forecast en route was an extensive squall line. As they approached the dark clouds, pilots heard other units abort; True, nicknamed "Fearless," later claimed not to have heard the order to return to Dobodura. Tatelman remembered: "I never believed that, and neither did anyone else. We all heard the order, but we were young and tough. I liked Colonel True because he was aggressive and wanted to fight. I wouldn't have liked following someone who wasn't like that."

The bombers were forced to fly so low by the weather that pilots had to open their windows to see the ocean below. When they broke out of the cloud, they were alone. Tatelman stated: "Off Cape Gazelle, we were jumped by 15 Zekes. They made 21 passes, but mostly broke off before they got into range. Of the ten that came in close, our gunners shot down three. Once we were out to sea after the attack, they left us alone."

B-25D gunships of the 500th BS, 345th BG, strafe the Japanese air base at Wewak, on the northern coast of New Guinea in 1944. (NARA)

"Bats Outta Hell" B-25J gunship 44-29600, flown by Second Lieutenant Francis A. Thompson, pulls up from a strafing run on a Japanese C-Type Kaibokan (ocean defense ship) in the Taiwan Strait, April 6, 1945. (NARA)

The 345th shot down 22 defenders and five probables, with many more destroyed by strafing, for the loss of two Mitchells. Colonel Smith of the First Air Task Force did not believe True and wanted to charge him with disobeying orders. When True reported to Gen George Kenney in Brisbane, he saw a headline in the local paper: "MacArthur Using Daring New Tactics, Sends B-25s Over Rabaul Unescorted." Kenney did not believe him either, but, in view of the success, a court-martial was deemed inappropriate. Instead, True was awarded the Distinguished Flying Cross (DFC) and the "Air Apaches" received the first of an eventual four Distinguished Unit Citations.

Armorers load ammunition for the eight .50 cal machine guns in the nose of this B-25J. (Key Collection)

Flying Tiger ace R. T. Smith in the cockpit of B-25H *Barbie III*, named for his wife, leads a formation of 1st ACG Mitchell gunships. (Key Collection)

The First "Wild Weasel"

During his second tour in 1945, Tatelman flew a war-weary B-25D Mitchell fitted with an eight-gun B-25J nose named *Dirty Dora II*, using special equipment to detect or "sniff out" Japanese early-warning radar sites, which were then destroyed by strafing. The experiment was highly successful, and *Dirty Dora II* could be thought of as the first "Wild Weasel" (modern USAF code for anti-radar aircraft/operations). By August 1945, he had flown a total of 100 missions as a strafer, a record unmatched by any other pilot. Gun-nosed B-25s were also operated in the Solomons by the 13th Air Force and in Burma by the 1st Air Commando Group in support of the Chindits, led by American Volunteer Group ace R. T. Smith.

My Personal War – Vic Tatelman

I knew not to do it… volunteer, that is. Never volunteer in the military was the universal axiom, but before I knew what I was doing, I raised my hand. It was procedural: After 50 combat missions in the Southwest Pacific Area, flight crews were rotated back to the ZI, the Zone of Interior, the US of A. At that time, it was early 1944, I had flown my 51st mission and was contemplating whether to stay on for a few more, when it happened: The Operations Officer at one of the briefings, with a paper in his hand, asked: "Anybody have any engineering training in college?" I don't know if it was the early hour or that I hadn't had my coffee yet, but that's when it happened, I raised my hand.

In 1941, before Pearl Harbor, one must have had least at least two years of college before the Army Air Corps would consider an application for flight school. I had completed two years of engineering in June of 1941 and immediately applied for pilot training, but it was October of 1941 before I received that magic letter.

In those days before the country was officially in the war, military flight schools, both Army and Navy, were modeled more or less on the standards of the Military and Naval Academies,

A trials B-25G gunship in Caribbean antisubmarine camouflage; Olive Drab over white. The 75mm cannon carried by this Mitchell could hole any German U-boat easily. (Key Collection)

lots of math and science and an Engineering "background" made life easier in ground school. So, on that morning in 1944 when the Ops Officer asked the question, I unthinkingly raised my hand. Since I had made the decision to stay on in the squadron for a while, I didn't think any more about it until one day, several weeks later, I was ordered to report to the CO, the Squadron Commander. Our CO at that time was Julian Baird, the best all-round CO in my entire military career. Looking at the orders he was holding, he said, "You're either in one helluva lot of hot water or somebody up top wants to talk face to face instead of communicating through channels. You're to report to Room such 'n such in the Pentagon in one week."

Well, that was kind of exciting – strange but exciting. I couldn't imagine what this was all about.

In those days, getting from New Guinea to the Pentagon was no small feat, especially in a week. Despite the fact that I had no clean Class A uniform (in New Guinea for a year, nothing was clean) I managed to appear halfway decently dressed in that room in the Pentagon within a week.

Evidently, my background had been checked and the powers-that-be decided that I wouldn't spill the beans to the Japanese, so I was briefed on the highly secret, at that time, Air Force (the designation had just been changed from the Army Air Corps to the Army Air Force) development program to nullify the accuracy of the German radar-controlled antiaircraft (AA) guns, the Wurtzberg Radar. Our bomb group tactics against the Japanese were primarily low-level, tree-top attacks where AA fire from large-calibre guns, radar-controlled or not, was fairly inaccurate. The enemy defence was primarily small-arms ground fire and it could sometimes be deadly.

But the heavy bombers, the B-24s, flew at medium altitude, 8,000 to 12,000ft [2,438 to 3,658m], where they encountered heavy AA over the targets, but from what we understood, the accuracy

Colonel True led the first Rabaul Raid in B-25D *Red Wrath* of the 398th BS. Here, the well-worn B-25D gunship displays the remarkable nose art of the "Falcons." (NARA)

was spotty. The intelligence people thought they had information that the Germans had given the Wurtzberg Radar know-how to the Japanese. That's where I would come in.

My job was to learn all there was to know about countermeasures developed to nullify radar accuracy, go back to the theater, and brief the combat crews on such techniques. I was sent to the various technical centers around the country: Wright-Patterson Field in Ohio; Eglin Field, the Naval Training Center in Orlando; Boca Raton Radar School in Florida; and Bell Labs in New Jersey to learn about that newly developed marvel called radar and the various techniques to counter it.

After about two months of great Stateside duty, I was headed back to the Southwest Pacific and was assigned to Section 22, the Intelligence Department of Gen MacArthur's Headquarters in Hollandia, Dutch New Guinea. Here I was given a map showing the location of the various heavy bombardment squadrons and a jeep to get to them.

The next month or so found me traipsing around New Guinea, then Leyte in the Philippines, talking to B-24 crews. I learned that the accuracy of Japanese heavy AA artillery hadn't drastically improved. Heretofore, most missions found the Japanese defenses somewhat delayed. Usually, the first elements of the attacking formation found AA fire light or even absent; it was the later part of the formation that received the heavy AA fire. Lately, however, returning crews were reporting heavy AA defenses even before reaching the target.

That meant the Japanese had developed an effective Early Warning Radar; to me, that was ominous. I reported that fact to my superiors at Section 22 and presented a suggestion. Instead of my flitting around the theater talking, let me design an airplane with radar homing gear, and actually home in on those new early warning radars and destroy them, instead of, or in addition

A B-25D of the 398th BS "Falcons," 345th BG, executes a low-level strafing attack on a Japanese base in New Guinea,1944. The lack of altitude is particularly obvious here. (NARA)

B-25D-20 41-30669 *Tondelayo* of the 500th BS "Rough Raiders," 345th BG (5th Air Force). Note the four-gun strafing fit in the nose, and the overpainted glazing above. (NARA)

B-25s attacking at low level drop "parafrag" fragmentation bombs. The parachutes slowed the bomb's fall sufficiently enough to allow the bomber to escape the explosions when they struck ground. (NARA)

to, trying to thwart them electronically. During my learning sessions in the States, I was shown an experimental radar-homing device being developed at Bell Labs in New Jersey. I suggested to the Captain that I reported to, to get that equipment over here, let me have an airplane in which to install it and turn me loose. My Captain went to the Section CO and in two days I had my orders. I could have any airplane in the theater, carte blanche at the Air Depot for modifications as I saw fit and arrangements for attachment to a combat squadron for rations, quarters and aircraft maintenance.

Since I had recently come out of a combat squadron, the 499th Squadron of the 345th Bomb Group, the famous, even then, Air Apaches, I opted to be attached to that Squadron. Too, I knew those people – they were my friends. A fact that hadn't crossed my mind was that all the combat crews, the people that I had flown with, had completed their missions and returned to the States. When I got to the squadron, the only remaining people I knew were the Squadron Section Heads, the ground officers who headed up the various sections, Communications, Engineering, Administration, Armament, etc. So, I moved in with them. Today, after more than 58 years, we still are as close as we were then.

Since I had flown combat in B-25s I naturally wanted one for my new project. There were many improvements in the development of North American's Mitchell Bomber, and those arriving in the theater now were the latest, the brand new "J" Model. Those had an eight-gun nose and a pair of "package" guns on each side of the fuselage, but the top turret had been moved forward into what was the navigator's compartment in the "D" Model, and I needed that navigation table for mounting the soon-to-arrive homing device. A search of the salvage areas turned up a war-weary

The famous B-25C strafer *Dirty Dora*, of the 499th BS "Bats Outta Hell," with myriad mission marks and garish nose art. Note the scabbed gun pack on the port fuselage.

"D" Model, a local engineering squadron performed a quick miracle, and I was soon flying it to the Air Depot at Biak, an island off the northwest tip of New Guinea. The Air Depot people had already received word that the aircraft I brought to them was to be modified as per my instructions, but they couldn't get to me for about ten days.

What the hell, since I was that close to Sydney and I had my own airplane and had ten days to wait and I hadn't had a leave in a year, I typed out a set of leave orders, signed General Southerland's name (forged, I should say) and flew down to Sydney. Sydney, even today my pulse quickens at the memory. What a paradise that city was in those days; clean sheets in a fine hotel, hot water in the shower, steak and eggs for breakfast, wine with dinner and girls…

Anyway, back to the war! Back in Biak, I had the airplane modified for single pilot operation, had a "J" Model eight-gun nose installed, moved the inverter switch that controls the power to the instruments up to the instrument panel, relocated the bomb bay door lever to the left side wall of the cockpit and most importantly, had the radar equipment installed where I had planned, on the old navigator's table. Not only did the powers-that-be send the equipment, but they sent a technician from Bell Labs to install and test it. We flew several local flights against our own radars in the area where it was tested and adjusted, tested and adjusted, again and again, until we were satisfied that it worked. The whole mission concept depended on this phase; I was overjoyed that it worked so well.

The technician that came with the equipment talked somebody in authority (or perhaps it came from Washington) to let him stay in the theater and fly with me to operate the device when we returned to combat operations. That in itself was amazing; as far as I knew, the guy was a civilian, but he turned out to be a genius with that equipment. And so, we returned to the old

The *Dirty Dora* groundcrew pose in front of their charge at Nadzab, New Guinea, February 1944. L to R: Herman Quinius, mechanic; William Kemmerling, armorer; Vic Bright, crew chief; Perry Scott, assistant crew chief.

squadron, by that time based at San Marcelino, in Luzon. I was assigned my old groundcrew, who promptly named the airplane *Dirty Dora II* and painted the Squadron's distinctive bat emblem on the nose.

Our targets and mission assignments with an all-volunteer crew came almost immediately, and we were extremely effective. The first three weeks in operation, we photographed and destroyed eight Japanese early warning radars. And it was exciting, without fighter cover, operating alone in enemy-held territory. Aside from some small-arms ground fire and being jumped once by a lone Japanese fighter, we were unscathed. Lawrence J. Hickey writes about the operation in his book, *Warpath Across the Pacific.*

In later conflicts, Korea and Vietnam, organized units came into existence specifically designed to suppress enemy radar – they were called "Wild Weasels". I guess I was the first Wild Weasel. After several weeks, our targets became fewer and fewer. The heavy bomber crews reported lower early warning Japanese defense activity and with that report, I felt a sense of accomplishment. At least, the results of my personal war were justified for the time and effort involved. My superiors at Section 22 evidently thought so too; I was recommended for a second DFC.

Everybody could see the handwriting on the wall – the Philippines had been retaken, the Okinawa operation had been bloody, but successful, and Japan itself was being destroyed, not only by the 5th, 7th, and 13th Air Forces on Okinawa and Ie Shima, but devastatingly, by the 20th Air Force, the B-29s from Guam and Tinian. So, I decided that radar-busting was

Dirty Dora was a hand-me-down veteran of the Battle of the Bismarck Sea. Flown originally by the 38th BG, it was one of the first B-25 gunships created by "Pappy" Gunn. Here, the aircraft is seen before conversion as a strafer. Note the .30 cal "cheek" gun in the bombardier's nose compartment.

becoming boring and applied to rejoin my squadron (instead of merely being attached). And in due course, I and my airplane, *Dirty Dora II*, became part of the famous (or was it infamous?) Bats Outta Hell, the 499th Bomb Squadron, once again, and finished out the rest of the war as a Flight Leader.

The Lady in Action
Vic Tatelman continued:

Dirty Dora II was initially ready for operations at San Marcellino, on the west coast of Luzon in the Philippines in February 1945. I had brought the plane up from the Air Depot at Biak, where the modification took place after my headquarters at Tacloban had arranged for rations, quarters and aircraft maintenance for me with my old combat squadron, the 499th of the Air Apaches, the 345th BG.

Frag orders (orders from authorizing headquarters outlining the strike, including target, enemy resistance expected; these were issued daily) from either Air Force Headquarters (ADVON – Fifth Air Force Advanced Echelon) or 5th Bomber Command, would specify an area of enemy territory to be covered. The area was determined from reports of bomber crews who suspected the presence of Japanese early warning radar. The Squadron Operations Officer would then select a volunteer crew and, depending on weather, we would search the suspected location using our radar-homing gear.

Engine maintenance is carried out on *Blondie's Vengeance*, a 7th Air Force B-25G gunship on Makin Island, 1944. The rough-and-ready conditions for crews and aircraft alike are clear.

There were always enough volunteers but after the first couple of missions, the crew stabilized into a more or less consistent group. The copilot, Byron Reed, however, was assigned specifically to me and we flew together throughout the rest of the war. We became close friends.

My assigned unit, Section 22, G2 (Ground Intelligence Section) of General MacArthur's headquarters was headed by General Walton; he briefed me personally at the time the whole scheme was developed three months previously, and especially admonished me to thoroughly photograph the sites of the Japanese radars before destroying them. The intelligence people desperately wanted those photographs to determine types of radar, the shape and configuration of the antennas, from which frequencies and range could be determined, and whether they were of German design. So, our aircraft-mounted cameras were as important as the homing gear. After several RCM (Radar Countermeasure) missions, we established our procedures and tactics, including the best times for attack, the most appropriate approach for photography and how to suppress ground fire. The photos we had taken began to appear on posters mounted in combat squadrons ready rooms, and I began to receive congratulatory comments from higher headquarters.

The Allied advance up the island of Luzon continued slowly. When Manila was taken and then the old Air Corps Base at Clark Field, we moved into the old Fort Stotzenberg at Clark. Our missions continued, we covered the whole area, even as far as the coast of China, but especially into northern Luzon and the adjacent island of Samar. On one such mission to

Vic Tatelman in the cockpit of B-25D *Dirty Dora II*: "I had to use an old B-25D because we needed the navigator's position right behind the cockpit for the electronic gear we carried to search for Japanese radars." This gunship was arguably the first "Wild Weasel."

northern Luzon, a Japanese radar was suspected at Lingayan on the north coast and our route to that area took us over the infantry advance line. As we flew over, I received a radio call from the ground troop commander asking if I could help him with a Japanese tank concealed in what looked like a barn, which was holding up his advance. Of course, I agreed and circled the barn, so as to positively identify the right target (all those barns looked alike to me). After I was assured from the ground, I set the barn on fire with machine gun fire, which silenced the tank and brought the Japanese tank crew running. With a "thanks" from the ground, I circled the area one more time (what's that line about curiosity killing the cat?), though this time too low. I saw it briefly before the strike; the ground radio vehicle had a vertical whip antenna which must have extended a hundred feet into the air. I couldn't have been that low and I saw it too late. I pulled the airplane up the instant I realized what it was, but it wasn't enough. We struck the wire rod at the wing root, missing the prop by inches. Furious with myself, I called off the RCM mission and returned to base expecting catastrophic damage to the airplane.

After examination, the crew chief wasn't too concerned; the airplane was repaired and ready for operations again in a week. I tried to identify the infantry unit with which we were involved to send my apologies for destroying their communications, but it just wasn't possible. I suppose their advance up the island was halted as much from my curiosity as from the Japanese tank, at least 'til they could replace the antenna. I excuse myself a bit because it isn't often Air Force people get to see real infantry ground action up close and I was fascinated.

NAA's period advertising for its Strafer B-25; Undoubtedly just as much for propaganda purposes, the artwork was far from misleading and echoed the aircraft's powerful capabilities well.

PAUL IRVIN "PAPPY" GUNN

Born in Quitman, Arkansas, in 1899, Paul Irvin Gunn fell in love with flying at the age of ten, when he saw his first aircraft. Leaving school after the 6th grade, he enlisted in the US Navy in 1918, where he trained as an aircraft mechanic. Gunn took flying lessons on his own time, and then reenlisted in 1923, graduating as an enlisted naval aviation pilot in 1925. Through most of the 1920s, Gunn served in VF-1 and as a flight instructor at Pensacola, then as a seaplane pilot and finally as a VIP pilot at Anacostia Naval Air Station in Washington, DC.

Retiring from the Navy in 1937, he worked for Bob Tyce, who had founded Hawaiian Air Lines, before moving to the Philippines in 1939, where he became the personal pilot of a wealthy Filipino. Gunn soon encouraged his employer to start Philippine Air Lines. With his extensive flight experience, he was given a wartime commission as a captain in the Army Air Forces following the outbreak of war in December 1941. He was forced to leave his wife and four children in the Philippines, where they were interned at Santo Tomas University. Following the success of his gunship conversions, Gunn was sent to Washington to explain the modifications. Told by USAAF engineering officers that mounting eight .50 cal machine guns in the nose of a B-25 was impossible, Gunn responded: "Look, I'm not an engineer. I'm not even an educated man. But what we have done works."

Gunn tested his own experiments by flying them in combat, where he was twice awarded the DFC, Silver Star, Legion of Merit, Air Medal, and nine Purple Hearts. Following the end of the war, Gunn rebuilt Philippine Air Lines, but died in a plane crash during a storm on October 11, 1957.

(Nat Gunn via John Bruning)

VICTOR TATELMAN

Born in Indiana in 1920, Vic Tatelman learned to fly at the age of 16. Joining the USAAF shortly before Pearl Harbor, he graduated in July 1942 as part of Class 42F from flight school in Stockton, California. He became an original member of the 499th BS "Bats Outta Hell" of the 345th BG. In 119 missions between September 1943 and July 1945 (over two full combat tours), he was awarded two DFCs, the first for leading a mission in New Guinea and the second for his work in developing the capability to detect and destroy Japanese radar sites; he also earned a Purple Heart and numerous Air Medals as the only strafer pilot to fly two tours.

On August 19, 1945, the Japanese surrender delegation returned to Ie Shima from Manila with two sets of the surrender documents. The first of two Mitsubishi G4M Betty bombers – carrying half the delegation and one of the documents – crashed in the Inland Sea when it ran out of fuel on the return to Tokyo. Tatelman, who volunteered to fly as copilot in Maj Wendell Decker's B-25J *Betty's Dream*, helped to escort the only set of surrender documents safely to Japan.

Tatelman transferred to the Air Force reserve while he finished a degree in aeronautical engineering. He was recalled to active duty in 1951 and flew F-80 and F-86 fighter bombers in Korea with the 35th Fighter-Bomber Wing. After retiring from the Air Force as a lieutenant colonel, Tatelman remained active in aviation and built several helicopters. He was the last B-25 combat pilot to fly a B-25, flying the restored warbird *Barbie III* at airshows across the United States, until he gave up night flying at the age of 85. Vic died peacefully in Winter Harbor, Florida, in December 2016.

(Key Publishing is grateful to the Tatelman Archive to be able to include anecdotes from Vic's extensive collection of reminiscences.)

Chapter 5
The Real *Catch-22*
Battle of the Brenner Pass

Unusually, the 57th BW and its fliers are better known thanks to fiction, rather than official archives and histories. Participating in this campaign was 21-year-old bombardier and aspiring writer Joseph Heller, who later used his experiences in Corsica and Italy to create *Catch-22*, a war novel so enduring in popularity that its title is now used generically to describe an impossible situation – between a rock and a hard place, so to speak. While Heller has always claimed the tale was a complete work of fiction, historians have demonstrated the characters were based on real people, with the events portrayed being fictionalized versions of actual occurrences. However, the real story of the wing's three B-25-equipped groups is far more interesting than Heller's transcending fictional tale.

The most famous member of the 57th BW was 486th BS bombardier First Lieutenant Joseph Heller, who flew 60 missions between June 1944 and January 1945, and turned his experience into the most famous war novel of the 20th Century, *Catch-22*. (Burton Blume collection)

B-25Js 9M *Daisy C* and 9V *Miss Rebel* of the 489th BS, 340th BG, negotiate the Brenner Pass during the winter of 1944–45, the worst European cold spell in a century. (All images in this chapter are from NARA unless otherwise stated)

Invading North Africa

The 310th, 321st and 340th BGs were formed in the summer of 1942 at Charleston Army Air Field in South Carolina, which was also the main aircrew training base for B-25 crews. Tutoring was rapid, with the North African invasion looming, but the crews were ready that autumn. The 310th landed in Morocco in late November, with the 321st arriving in late-January 1943 and the 340th following in March.

The 57th BW was commanded by Brig Gen Robert D. Knapp, a World War One aviator who had been put in charge of all B-25 training in 1941. As a 43-year-old colonel, he convinced his superiors to allow him to take the 321st into combat. Over the course of the next year, the Mitchell groups fought through the North African campaign, the Sicilian invasion and beginning of the Italian "push." During this period, Knapp would be found in the left-hand seat of his B-25 on every tough mission, before he was promoted and placed in command of all three groups in January 1944. Portrayed in the novel as "Gen Dreedle," he was the polar opposite of his literary doppelganger.

Left: Technical Sergeant Fred Lawrence was the crew chief responsible for B-25J-1 43-27698 *Peggy Lou*'s remarkable operational record.

Below: B-25J *Peggy Lou* arrived at the 445th BS in May 1944. By the end of the war a year later, it carried 137 mission symbols on its nose. The crew never aborted a mission for mechanical failure, and its nickname in the squadron was "Lucky 13." (Fred Lawrence collection)

Helbig's Raid

The 340th BG was the victim of bad luck on three occasions. Shortly after being organized, it suffered the loss of 18 B-25s in South Carolina due to a freak hailstorm. On March 18, 1944, when based at Pompeii Airfield, south of Naples, the group lost all 88 B-25C/Ds when Mount Vesuvius erupted for the first time in 175 years. The volcano visited more destruction on the USAAF in Italy than the Germans had accomplished in 18 months of combat. Reequipped with brand-new B-25Js by mid-April, the unit arrived at Alesani Airfield on Corsica at the end of the month. On May 13, legendary Luftwaffe bomber leader Oberst Joachim Helbig led 90 Ju 88s of I and II Gruppe, Lehrgeschwader I, and 60 Ju 88s from Kampfgeschwader 76 on what would be the Luftwaffe's last bombing raid in the Mediterranean Theater. The target was Alesani Airfield, as one pilot in the 488th BS recalled: "We had all these nice new shiny silver ships, and they reflected the light from the fires so well that the Germans had no trouble spotting where to drop their bombs." The 340th BG lost 65 B-25s in total, with just 20 left in flyable condition. On May 15, in a gesture of defiance, those surviving Mitchells attacked the railroad tunnel near Itri, Italy, with good results.

Second Lieutenant Joseph Heller arrived at Alesani as a replacement crewman on May 18, and was assigned to the group's 488th BS. The week before, while awaiting assignment at the replacement center in North Africa, he had written in his diary he was ready to see action: "I want to see skies full of flak, and fighters screaming past in life and death duels high in the clouds." Four days after his arrival, he flew his first mission.

Helbig's raid would be memorialized by the 57th BW for the rest of the war by the appearance of the surviving Mitchells. In the weeks following the event, B-25s in the wing's 12 squadrons were given a hasty coat of camouflage on their upper surfaces, with no two alike. No one ever recorded which hue was used, but color photos indicate the most likely source of the paint was French and British, cadged from Free French and RAF squadrons on Corsica, and being those colors closest to Olive Drab.

Volcanic ash covered all 88 B-25C/D Mitchells of the 340th BG following the eruption of Mount Vesuvius on March 18, 1944. The fabric has been burnt away from the elevators on this aircraft. (USAAF via Setzer)

A B-25D of the 321st BG flies over Mount Vesuvius following the March 18, 1944, eruption. The natural disaster did more damage to the USAAF in Italy than the Luftwaffe. (USAAF via Setzer)

The paint was applied with mops and brushes, without primer, and weathered badly under the hot Corsican sun and frequent rainstorms. While the average operational life of a B-25 on Corsica was 20–25 missions, before it was either shot down or damaged so badly by German flak over Italian targets it ended on the airfield junk pile, several Mitchells survived more than 100 missions.

"USS *Corsica*"

The island of Corsica, located approximately 100 miles (161km) from the western coast of Italy and an equal distance south of France, became known as "USS *Corsica*" for the many airfields built on it over the spring of 1944, which turned it into an "unsinkable aircraft carrier."

Its location enabled fighters and bombers to provide air support to Allied operations throughout Italy and the south of France. Ghisonaccia Gare was the largest and busiest airfield. The home base of the 310th BG, it also hosted the USAAF's 57th, 324th, and 350th Fighter Groups (operating P-47 fighter bombers), Free French and RAF units. The 321st BG was based at Solenzara Airfield, while the 340th continued to fly from Alesani.

Each group had four 16-aircraft squadrons. Bombing missions generally involved 16–20 Mitchells, with the formation comprising aircraft from two squadrons. Each group usually flew two separate morning and two separate afternoon missions, weather permitting, and crews could generally expect to fly a mission every other day. After the Battle of Cassino, targets in Italy were predominantly road and rail lines, bridges and other transportation infrastructure, as well as German troop concentrations, storage sites and convoys.

Italian Campaign

The war in Italy was one that Winston Churchill alone had wanted to fight. Following the invasion of Sicily, which secured Allied control of the Mediterranean, American military leaders wanted to withdraw US forces to Britain, in preparation for the cross-channel invasion of France. Churchill, who did not want to confront the Germans on the old Western Front (and run the risk of a second national bloodbath in 20 years), argued the Allies should strike "the soft underbelly" of Europe, invading Italy and advancing along the peninsula. When the Americans resisted the idea, he argued such an invasion should take place if only to occupy the Germans and keep them from getting any "breathing room" over the ten months before Normandy. American agreement was finally given, albeit reluctantly.

The subsequent Italian campaign was marked throughout by poor planning and generalship on the part of top Allied commanders. The Germans planned originally to evacuate most of the peninsula and make a stand on a heavily fortified line in the Apennines in the north, designed by Generalfeldmarschall Erwin Rommel. However, the equal-ranking Albert Kesselring convinced Hitler to reinforce in the south, just one month before the Allied invasion at Salerno in September. Despite this, the Germans came close to expelling the invading troops from their beachheads.

Italy's "soft underbelly" turned out to be nothing of the sort, instead presenting one fortified mountain after another, with the Germans mounting effective defenses on succeeding river lines, until they made their winter stand at Cassino, a third of the way between Naples and Rome. Over the course of four battles, fought from January–May 1944, the Allies were finally able to take Cassino, but only

B-25J of the 310th BG damaged during the Leghorn mission of June 22, 1944. Tail-gunner T/Sgt Jerry Campbell wrenched his back when he fell 10ft to the ground, attempting to exit through the emergency hatch. (George Underwood)

Running the gauntlet: B-25J *8H* of the 488th BS, 340th BG, goes down on fire after being hit by flak over the Brenner Pass, January 1945. (USAAF via Setzer)

after the battlefield had been cut off from regular supply for the Germans by Allied air power, including the 57th BW's Mitchells. In the aftermath of the German loss at Cassino, there was a real chance of rolling up the German Tenth Army in an Anglo-American pincer movement south of Rome, with a surrender that would have ended the war in Italy and made possible Churchill's dream of a direct thrust into Eastern Europe from northern Italy. However, US Fifth Army commander Lieutenant General Mark Clark's fixation on becoming the "Liberator of Rome" allowed the Germans to escape. The summer of 1944 was then spent chasing German units as they made a fighting retreat to Rommel's "Gothic Line," which they reached in late August. Compounding these issues, Allied manpower was diverted to the invasion of Southern France (Operation *Dragoon*) that very month.

The Gothic Line was a formidable barrier. Extending across Italy along the line of the Apennines, it featured 2,376 pillboxes with interlocking fire, 479 anti-tank, mortar and assault gun positions, 73½ miles (118.8km) of barbed wire, and several miles of anti-tank ditches. All fortifications were located on steep hillsides, which reduced combat to small-unit infantry engagements.

Operation *Olive*, the Allied offensive to break through the Gothic Line, began in late August and quickly became the biggest battle ever fought in Italy, with more than 1,200,000 men involved. The B-25s of the 57th BW were involved intimately, as they targeted German transport routes. By mid-October, the armies involved had fought each other to a standstill. Attrition was so severe that infantry divisions throughout the British Army were reduced from four to three regiments, to make up for losses incurred in Normandy and Italy.

Battle of the Brenner Pass

At the end of October, the autumn rains turned to snow, heralding the arrival of the coldest European winter in a century. It was clear the Allied armies were not strong enough to overcome the

Personnel of the 57th BW spent 11 months living under canvas on Corsica, from their arrival at the end of April 1944 to their departure for the Italian mainland at the end of March 1945.

well-supplied Germans, who were receiving 24,000 tons of supplies and equipment – 600 percent of their minimum daily requirements. These were shipped from Munich and Augsburg by rail through the Brenner Pass to Bologna, the main German supply center, with less than 12 hours required for a successful journey. Now that large-scale offensive operations were impossible until the spring thaw, a way had to be found during the intervening winter months to reduce enemy strength. The target would be the Brenner Pass rail line, and the 57th BW was given the assignment of carrying out what was named Operation *Bingo*.

The first mission was flown on November 6, 1944. The 310th BG hit the transformers at San Ambroglio, while the 340th and 321st BGs targeted those at Trento and Ala, respectively. All missions were successful, and electrical power along the Brenner line was cut to trains as far north as Balzano. This forced the Germans to rely on coal-fired locomotives from elsewhere, as part of their overstretched rail transportation system.

The 321st BG was assigned a particularly tough mission on November 10. Forty-four Mitchells were to bomb a rail bridge and newly completed ferry terminal at Ostiglia. The bridge was defended fiercely, and four B-25s (including two formation leaders) were shot down, while 30 aircraft were holed by antiaircraft fire. Thirty-two aircrew were also wounded, making this one of the heaviest casualty lists of any mission; despite this, seven spans of the pontoon bridge were destroyed. Newcomer 2nd Lt Paul Young remembered the sortie (his second) well: "It was my first mission where the Germans made a serious attempt to kill me. It wasn't easy to sit there as copilot and take it as all that flak exploded around us, but in retrospect it was better I had that first experience as a copilot where I wasn't responsible for a crew."

Several B-25s of the 321st BG lumber along the taxiway at Amendola, Italy, during late 1943, including the veteran *OH*-7. A B-25C-1-NA, the aircraft's true serial number was 41-13207, but, in this image, the rudder of a different B-25 had been fitted, bearing the "last three" 486 (the serial number combination 113486 did not exist for any B-25s). (Malcolm V. Lowe Collection)

When a crewman panicked during a bombing mission, it was called "I was so scared I was all up inside my flak helmet." This amusing photo demonstrates the metaphor perfectly!

"Sunny" Italy

It might have been a constant refrain of the tourist brochures, but, during the winter of 1944/45, "sunny" Italy was anything but pleasant. The 57th BW's B-25s had just limited heating, while the oxygen systems had been removed due to the potential fire danger from flak hits. Radioman Staff Sergeant Jerry Rosenthal remembered:

> Anoxia (lack of oxygen) was a big problem, since the missions into the Brenner Pass were generally flown at 11,500 to 15,000ft [3,505 to 4,572m]. We would take our gloves off to check our fingernails for signs of anoxia even though we couldn't do anything about it. We also got to know all about aerotitis media, the inflammation of the inner ear from changes in altitude. The air temperature at my station as radioman was around 25° below [-32°C], and if you took off your glove and touched anything, you could freeze your skin to whatever it was. The pilots, the bombardier, and the turret gunner in the nose had a couple heaters there that probably got the temperature up to 10° below [-23°C], as did the tail gunner. But the radioman's position was unheated and drafty, since it was right behind the bomb bay.
>
> I acquired a blue wool electric flight suit, even though there was no electricity available for it in the airplane, but I wore it over my GI long johns, then a wool shirt over that followed by my A-2 jacket with a super heavy sheepskin coat over that! Wool socks, then GI shoes with the wonderful

OH-7 flew on more than 80 missions, but it had to be crash-landed due to combat damage on January 13, 1944, while being flown by Flight Officer Dean B. Vallery. (Malcolm V Lowe Collection)

sheep-lined flying boots over them. I had sheepskin overalls to cover the wool GI pants, and with all that, I was able to prevent frost bite!

In addition to the cold, there was nature to contend with. Turbulent winds swept down the pass at speeds of 50–60mph (80–96km/h). The mountains on either side were 8,000–9,000ft (2,438–2,743m) high, with several reaching 10,000ft (3,048m). One member of the 445th BS, copilot 2nd Lt Victor Hanson, remembered a mission in which his aircraft was hit and dropped out of formation. The pilot turned away from the flak into a box canyon, which was too narrow to turn around in:

> We were climbing against the downslope winds that were going down only a little bit slower than we were going up. We were holding maximum power and just barely climbing. All those rocks out in front definitely had my attention as we got closer and closer. In the end, we cleared the ridge by maybe 20ft. That was my scariest mission and it didn't have anything to do with the Germans.

Little Respite

There was one constant factor throughout the 57th BW's operations in Italy – the threat from enemy ground-to-air fire. Primary defense was provided by IV Flak Korps, which, by November 1, 1944, was operating 366 8.8cm antiaircraft guns, stationed from Verona to Innsbruck, and more were added every month. The flak map of 2nd Lt Young's bombardier in February 1945, which had flak positions shown as red dots, depicted a solid red line, half-an-inch wide, running between those two cities; there were 541 guns by March 1945.

B-25Js of the 321st BG unload their stores over a Brenner Pass target.

At Rovereto, known as the worst flak trap in Italy, the bombers could attack from just one direction. The German counter was to build concrete gun pits halfway up the mountains on each side of that narrow pass, almost at the bomber's altitude. German observers with theodolites were stationed on mountain tops where they could determine the bombers' altitudes within a matter of feet. According to Young: "If they had had proximity fuses, we could never have flown the missions."

The Squadron Leader

One of the 321st BG's best pilots was 1st Lt Dan Bowling, a miner's son raised in hardscrabble Bisbee, Arizona, who had learned at an early age never to back down from a fight and had a reputation for not "taking guff" from anyone, regardless of rank. He arrived at Corsica in late August 1944, three weeks before his 22nd birthday. Assigned to the 445th BS, the squadron check pilot told him on their third flight that Bowling had more hours as a B-25 first pilot than his training officer. Four missions later (all flown during his first week on the unit), Bowling was awarded "lead pilot" status. After flying several high-loss missions, Bowling and his bombardier, Joe Silnutz, decided to find a way to reduce casualties, yet still achieve an accurate drop. At the time, standard tactics involved aircraft flying straight-and-level approaches for several minutes prior to releasing bombs, which allowed German gunners time to fire several volleys at the bombers. Bowling recalled:

> Joe and I decided that the only way to survive our missions was by performing evasive action. We practiced many times on the bomb range. We would fly a certain compass heading to the practice target circle, then turn ten to fifteen degrees right or left, then change again to a different compass heading and immediately change to the exact course to target. Joe's timing with that Norden sight was so accurate that we would only be thirty to forty seconds straight and level to the target.

Shortly after they devised their evasion tactic, Bowling was made Squadron Lead Pilot and Silnutz, Squadron Lead Bombardier:

> When I was out front ahead of everyone else, they had to do what I did, so they followed me. I could look out just after we changed course and see a barrage of flak go off right where we would have been had we continued. Then we'd turn and there would be another barrage go off where we would have been. When we turned on to the bomb run, the gunners were so confused they didn't have the time to put up that last volley before we dropped and broke formation.

Just a few other lead crew pilot-bombardier teams tried to emulate their bombing tactic. Colonel Smith, the 321st BG Commander, ordered crews to fly straight and level for four minutes since he was quite open that "I want a star when I leave" and held bombing accuracy as the key to that. Bowling told the pilots to follow him: "I was proud of two things about the missions I led. One was that we had the highest bombing accuracy of anybody in the group, and the other was that I had the lowest losses. We got the target and we didn't lose our friends." Eventually, the 445th Squadron Commander, Col Cassidy, ceased arguing with Bowling. Instead, he received the highest unspoken praise possible: he was the pilot picked to lead every tough mission during the worst period of the Battle of the Brenner Pass. As 2nd Lt Paul Young put it: "Dan was the squadron leader. That's different from the squadron commander."

Catch-22

Originally, an operational tour for crews in the 12th Air Force, of which the 57th BW was part, was 50 missions. After the failure to "bag" the German Army in June 1944, the Italian campaign quickly became a secondary front; following the Normandy invasion, units fighting in Western Europe had priority for personnel and equipment.

With aircrew replacements to the wing few and far between, the mission total for a tour changed, as Army leadership looked to a final victory. Upped to 60 in November, it then rose further, to 65, in December, and 70 in January 1945. By late February, crews were informed that their tour of duty was now "for the duration," as in until the war was won in Europe. Some tried to turn in their wings,

The mission to the Avignon Bridge on August 18, 1944, was memorialized in film *Catch-22* as the "help the bombardier" scene. Joseph Heller considered it the most terrifying mission he flew.

willing to accept transfer to the infantry as an escape from repeated "charges of the light brigade" into the Brenner Pass, but they were refused. Paul Young remembered: "It really was like what was in the novel: you had to be crazy to continue, but if you tried to get out, that meant you were sane and you had to stay. It wasn't called Catch-22 or anything, but the policy was there."

On August 18, 1944, the 321st BG flew one of its most dangerous missions to attack the French battleship *Strasbourg* and cruiser *La Gallisoniere* in Toulon harbor, to prevent their use against the Allies in the invasion of Southern France. A total of 36 B-25Js dropped with 100 percent accuracy, although 27 bombers were hit by flak in the process. (USAAF via Setzer)

Victory

The war's finale prompted the 57th BW to leave Corsica for bases on the Italian mainland, closer to the front, in early April 1945. The campaign had achieved success; at the end of March, the quartermaster general of the German 14th Army reported the Brenner line was open for just 12 days that month, and that it was broken permanently in four places. Crucially, this meant cargos had to be unloaded and reloaded at each break in the line, and shipments took 7–8 days to get from Munich to Bologna, far slower than the original 12 hours. By this time, less than 20 percent of the minimum needed supplies were getting through.

The final Allied offensive in Italy commenced on April 9, 1945, just as spring came to the Italian Alps. The 57th BW Mitchells flew a "maximum effort" in support of ground units, with each group and squadron conducting multiple sorties. The 340th BG's war diarist, 1st Lt Glenn Pierre, wrote of the day: "A terrific air pounding of German positions northeast and east of Bologna, by 600 bombers and hundreds of fighters starting shortly after one o'clock. 340th Group put up 76 aircraft, almost half the B-25 wing effort. Our targets were two artillery concentration areas near Imola. Photo Interpreter says most of the 13 boxes of six aircraft each bombed very accurately."

Perhaps predictably, given the massive air bombardment, the end came quickly with German armies cut off from their supplies, primarily due to the successes of the Battle of the Brenner Pass. On April 18, US Eighth Army units in the east broke through the Argenta Gap and sent armor forward to encircle the German Tenth Army and join with the American IV Corps, which had advanced from the Apennines in central Italy. The remaining defenders of Bologna were now trapped. 1st Lt Pierre wrote, "Bologna has fallen to the Fifth and Eighth armies, it was announced today. Air support undoubtedly was a big key in making the Germans release their iron grip of last autumn and winter. With their supplies cut off by bombing and communications badly slashed, the Germans could only fall back under the tremendous armored drive of our Allied 15th Army Group."

Dan Bowling of the 445th BS, 321st BG, in *Peggy Lou* drops fragmentation bombs over antiaircraft sites in the Brenner Pass.

Captain Tim Jackson, of the 445th BS, wrote: "Going home is the topic of the day. Everyone constantly talks or makes a ready listener for such talk. The war from this group's point of view is over. Targets change so fast that missions are uncertain from day to day." As much as everyone expected the end, there was still time for fighting and dying. The 321st BG flew its 895th mission on April 25, to bomb the Cavarzere road bridge, led by Capt Wayne Kendall of the 445th BS in *Spirit of Portchester*, which was a SHORAN (SHOrt-Range Navigation)-equipped B-25, with nine other Mitchells. These were from the 446th BS and included the veteran 43-4074/27, flown by 1st Lt Roland Jackson. Once again, the mission proved "hot," with German flak reaching for the aircraft as they approached the target. B-25 43-4074, which had survived the battle of Cassino, the invasion of Southern France, the first assault on the Gothic Line and the entire Battle of the Brenner Pass, was hit, causing an engine to fail and setting the aircraft alight. Gunner SSgt Joseph Dalpos, radioman Sgt Henry Nichols and tail-gunner Sgt George Darnielle baled out of the stricken bomber, but the latter's parachute failed to open, and he fell to his death – he would be the last member of the 57th BW killed in action. Jackson crash-landed at the first airfield he spotted, with the pilot and copilot managing to pull the wounded bombardier from the nose and escape the burning bomber.

By April 29, German Army Group C was in retreat on all fronts, having lost most of its combat strength. General Heinrich von Vietinghoff, who had stymied Allied armies from Salerno to the Gothic Line, was left with little option but surrender. Hostilities came to a formal end on May 2, 1945. Hearing news of the surrender, Jackson wrote:

> After almost two-and-a-half years of slugging, bombing, mud and mountains, the enemy collapsed practically overnight and the Italian campaign closed in a blast of superlatives. We've made greater gains than any other theater; we've taken the biggest bag of prisoners on all fronts; we're the first theater to wind up; and we're the first theater to receive an unconditional surrender

The Brenner Pass rail line in August 1944 – a challenging target for all the 57th BW B-25 crews.

from an army group. As for the group, we just wound up our busiest month in existence and set a lot of records for future operations to aim at.

During the Battle of the Brenner Pass, the 57th BW flew 6,839 individual sorties between November 6, 1944, and April 6, 1945, in approximately 380 missions. Forty-six B-25s were lost, while 532 were damaged, and more than 500 aircrew were killed or wounded.

Left: "Bubbletop" P-47D Thunderbolts of the 350th FG escort B-25Js of the 487th BS, 340th BG, including 43-27478 *Yahoudi* during a mission against the Brenner Pass rail line. (USAAF via Setzer)

Below: Bearing the Roman Numerals "III" on its vertical tail to denote the 447th BS, 321st BG, this B-25D was on a mission over Italy when photographed in this genuine contemporary color image. Officially, the aircraft was a B-25-D-30-NC, serial number 43-3522. (Malcolm V. Lowe Collection)

STERLING DITCHEY

First Lieutenant Sterling Ditchey navigated his B-25 12,000 miles (19,312km) from Florida to Corsica via the "Southern Route" – British Guyana, Brazil, Ascension Island, Liberia, Algeria, Corsica – using "four maps, a protractor, and four pencils." Ditchey flew 72 missions between May 1944 and February 1945 and came home six weeks before his 21st birthday.

DAN BOWLING

First Lieutenant Dan Bowling arrived in Corsica in August 1944, two weeks past his 22nd birthday. One of the most experienced pilots in the 445th BS/321st BG, he was eventually picked to lead all the squadron's difficult missions. Fellow pilot Paul Young described him: "Dan was our squadron leader. That's different from the squadron commander."

(Dan Bowling via Lance Bowling)

Chapter 6

Leatherneck Ops
Semper Fi Mitchells

Least-known of all B-25 variants, the PBJ-1 found its origin in an agreement made in mid-1942, between the US Navy and the USAAF. The bomber was the result of the Navy's desire to acquire a long-range, land-based maritime reconnaissance and patrol aircraft. The Air Force saw this as desire by the Navy to encroach on its land-based bomber primacy.

Resolution came through the USAAF's need of a plant to manufacture the new B-29 Superfortress. The Navy owned a plant at Renton, Washington, used by Boeing for production of the PBB-1 Sea Ranger. The USAAF proposed the Navy cancel that program and turn over the Renton factory for B-29 production. In exchange, the Air Force would drop its objections to the Navy's acquisition and operation of land-based bombers. The Air Force would transfer the B-24 Liberators, B-25 Mitchells, and B-34/B-37 Venturas then currently in use by the Air Force's Antisubmarine Warfare (ASW) squadrons and allow the right to acquire more; the Navy agreed readily, so the B-24 became the PB4Y-1, the B-25 the PBJ-1, and the Ventura the PV-1.

Following the agreement, 50 B-25Cs and 152 B-25Ds operated in the ASW role were transferred to the Navy with the designations PBJ-1C and PBJ-1D, respectively. The designation suffix of these and all subsequent PBJ-1 variants corresponded to that of the corresponding B-25 model. The aircraft carried Navy serial numbers beginning with 34998, and the first PBJ-1s arrived in February 1943.

Taken in the summer of 1944, this image captures a formation of VMB-433 PBJ-1Ds over the Pacific. (All images in this chapter are from NARA unless otherwise stated)

A US Marine Corps PBJ-1D bombs Rabaul, in late 1944. Note the prominent APS-3 radar unit in a dome where the ventral turret was installed originally.

Lumps and Bumps

Many PBJ-1C and D Mitchells were equipped with an APS-3 search radar inside a bulbous radome fitted to the bomber's nose. Alternatively, the radar was carried in a belly radome, where the lower turret had been fitted originally. The PBJ-1H and J mounted the APS-3 in a radome on the right wingtip. Toward the end of the war, the top turrets of several PBJ-1Js were removed to save weight, since the threat of interception by Japanese fighters had become relatively rare.

In 1944, PBJ-1H 43-4700 (BuNo 35277) was modified for aircraft carrier catapult launch and arrest. The first landings and catapult take-offs took place aboard USS *Shangri La* (CV-38) on November 15, 1944. While the experiment was successful, no further work on a carrier-based Mitchell took place, though these tests would later lead to development of the carrier-based AJ-1 Savage and A3D Skywarrior bombers in the postwar years.

During 1945, the Naval Ordnance Test Station at China Lake, California, carried out a joint program with the Harvey Machine Company to develop an automatic launcher for 5in (127mm) spin-stabilized rockets from the nose of a PBJ-1J. The rockets were carried in two rotating drums with five stores each, and they could be fired in salvoes of five or singly at 0.3 second intervals; a deflector tube diverted the exhaust blast downward at the aft end of the nose section. PBJ-1J 35849 (ex-USAAF 44-30980) was selected as a test aircraft for the project, but the trials failed to prove the concept sufficiently effective to warrant production.

The Mitchells were operated almost exclusively by the US Marine Corps, a little-known aspect of marine aviation during the war. In February 1943, the Marines began to receive its first PBJ-1C

VMB-613 moved from Iwo Jima to Okinawa in July 1945, where it trained to use the sizeable 11¾in Tiny Tim rocket for anti-shipping missions. Between August 9 and 15, the unit sank seven ships off Kyushu, the first operational deployment of this weapon. The bombers had all guns apart from those in the tail position removed, and they were painted overall dark blue for night sorties.

and PBJ-1D Mitchells. Marine Bomber Squadrons (VMBs) were established, beginning in March 1943, when VMB-413 was commissioned at Marine Corps Air Station (MCAS) Cherry Point, North Carolina. By the end of 1943, eight VMB units were equipped with PBJs, and Marine Medium Bombardment Group, VMB-61 was formed. Seven PBJ squadrons reached the Pacific before the war ended: VMB-413, -423, -433, -443, -611, -612 and -613. Four more squadrons were in the process of formation in late 1945, but these had not yet deployed by the time the war ended. While 173 men and 45 PBJs were lost in the Pacific, few people today are even aware that Marines flew the Mitchell bomber in World War Two.

Marine PBJs eventually operated from the Philippines, Saipan, Iwo Jima and Okinawa during the last year of the Pacific War. The primary mission was long-range interdiction of enemy shipping, and harassment of bypassed enemy island bases in the south and central Pacific. The PBJ-1J Mitchells that were the primary subtype were late-production aircraft fitted with four hardpoints in each outer wing to carry 5in (127mm) High-Velocity Air Rockets, which became the anti-shipping weapon of choice.

The field modified .50 cal waist gun position of an early production PBJ-1D. Note the battle damage at top right, and the baffle far left to reduce wind blast on the gunner.

Nocturnal Debut

The first PBJ squadron to enter combat was VMB-413 "The Flying Nightmares," beginning in March 1944. Commanded by Lt Col Andrew B. Galatian, Jr., the squadron operated from Stirling Island in the Treasury Islands group. The primary mission was heckling bypassed Japanese installations at Rabaul on New Britain and Kavieng on New Ireland.

Harassment missions were flown at night, regardless of the weather, and the heavy rains and strong winds of the spring monsoon buffeted the PBJs to and from their targets. After breaking through the foul weather, aircrews then had to contend with enemy searchlights, antiaircraft fire, and an occasional Japanese night fighter. Afterwards, the crews were forced to fight their way back through the inclement conditions. Losses were exceptionally heavy during the unit's first two months of operation, with five aircraft and 27 men lost, mostly to enemy action. One day, worried about losses, Lt Col Galatian asked his pilots if the missions should stop. Captain Robert Millington reportedly responded: "Sir, we are marines, and we don't quit!"

Despite the death and destruction, VMB-413 did not quit. The night heckling continued, although by May the number of combat-ready aircraft was reduced due to operational losses. On May 15, 1944, after 60 days in the combat zone, the unit was relieved by VMB-423 "The Seahorses" for two months of rest and recuperation (R&R) in Sydney, Australia, and Espiritu Santo, Vanuatu, south Pacific.

With the departure of "The Flying Nightmares," Lt Col John L. Winston's VMB-423 began operations, and the squadron was assigned a special mission just two weeks after its arrival. On May 27, 1944, a crew dropped a 65ft (20m) scroll on Rabaul, signed by 35,000 Oklahoma schoolchildren who had collected enough money for war bonds to buy a new bomber.

Then, VMB-423 moved to Green Island, located 60 miles (96.5km) east of New Ireland and 40 miles (64.3km) northeast of Buka, in June 1944, but within 14 days, two aircraft and crews were lost to the enemy and the elements. PBJ operations from Green Island doubled in July 1944 with the arrival of VMB-433 "The Devils," under the command of Maj John G. Adams.

Doolittle's Mitchells were not the only examples of the B-25 to grace the deck of an aircraft carrier. These anti-shipping radar-equipped US Marine Corps PBJ-1D bombers were in transit to the south Pacific when photographed aboard escort carrier USS *Natoma Bay* (CVE-62) during summer 1944. (John Batchelor Collection via Malcolm V. Lowe)

A well-weathered and radar-equipped PBJ-1D Mitchell of VMB-433 in flight over Emirau airfield, Papua.

Following the much-needed recuperative period, VMB-413 returned to combat in July, operating from Munda, New Georgia, in the Solomons. Galatian led "The Flying Nightmares" on its first daylight mission of the second combat deployment on July 29, 1944. Nearing the target, the Mitchells flew at just 150ft (46m), and the first run was made at 230mph (370km/h). The first section flew over Sipasai Island, where it dropped eight 100lb (45kg) general-purpose bombs near the middle of the island, then strafed a small auxiliary vessel spotted off the southeast coast. Galatian led five more runs before the unit's luck ran out as the Japanese defenders responded with blistering antiaircraft fire from the village; the PBJs took hits in their nose sections, engine cowlings, and wings.

On the seventh run, Capt Millington's aircraft, which was flying on the left of the first section, fell prey to an explosive 20mm round that hit the lower fixed nose gun, tearing both weapons from their mounts and putting them out of commission. The Plexiglass nose was shattered and the navigator, 1st Lt Joe Decuester, was knocked from his position. Millington was forced to ditch, and the crew was rescued five hours later.

Three-tone PBJ-1J gunships of VMB-433 bask in the harsh sun at Emirau airfield, 1944.

Island-Hopping

In August, VMB-413 and VMB-423 were transferred to Emirau Island, 250 miles (402km) south of Rabaul and 600 miles (965km) south of Truk, as part of Marine Air Group (MAG) 61. As these operations from Emirau began, Lt Col Dwight M. Guillotte's VMB-443 "Fork-Tailed Devils" arrived to participate in the neutralization of the Bismarck archipelago, in addition to heckling missions against Kavieng and Rabaul. The units would exchange their PBJ-1C and -1D Mitchells for rocket-armed PBJ-1Js in the latter part of 1944.

As the focus of Marine aviation continued in late 1944 with what was essentially "guard duty" against bypassed enemy bases, Maj Gen Ralph J. Mitchell, Commander of the 1st Marine Aircraft Wing (MAW), wanted to see at least some of his squadrons participate in the Philippines campaign. MAG-12's F4U-1D Corsairs, escorted by PBJs from MAG-61 to provide navigation, arrived at Tacloban airfield on Leyte in December 1944.

However, the PBJ units would not begin operations in the Philippines until March 1945. Until then, they continued to pound the Japanese in the Solomons, and MAG-61 would harass the Japanese at Rabaul until the closing days of World War Two. Writing at the end of the war, Capt E. J. Molloy of VMB-413 said of his unit:

> The mills of evaluation in war grind slow. History, an exacting mistress, will in her own good time assess what VMB-413 accomplished. We of the squadron cannot say. We can produce no heroes. We can sing of no glories. The nature of our mission did not lend itself to the spectacular. It was difficult, dangerous, disheartening – and routine.
>
> The bulk of our attacks against the enemy were at night. Occasionally a fire was started in Rabaul town; an ammunition dump touched off, a bullseye (sic) scored on a cluster of searchlights. Infrequently an enemy night fighter gave chase. What could not be observed was the effectiveness of that constant heckling over enemy territory. What percentage did we contribute toward the final neutralization of Rabaul in hours of sleep lost, in personnel killed or wounded, in repair and maintenance work interrupted? Someday, a line in a history book may give the answer.

VMB-611 "Black Seahorses" was commissioned on October 1, 1943, at Cherry Point, North Carolina, under the command of Lt Col George A. Sarles, an outstanding leader who had flown combat missions in the SBD Dauntless at Guadalcanal in 1942. Equipped originally with the PBJ-1D, the squadron

PBJ-1J gunships of VMB-433 at Emirau Island, autumn 1944. Conditions at locations such as this were austere and challenged air and groundcrews alike.

An atmospheric study of a PBJ-1D of VMB-433, releasing its bombload over Rabaul, New Britain Island, in 1944.

trained on the US East Coast until August 5, 1944, when it moved to the western Pacific. This journey would become an epic example of human survival, in the face of bureaucratic incompetence unlike any experienced by any other American unit during the war.

Half the aircrew flew 14 PBJ-1Ds to San Diego, where they were put aboard the escort carrier USS *Manila Bay* (CVE-68) for transport to Hawaii. Once there, they moved to MCAS Ewa. The ground echelon and the other half of the aircrews went aboard SS *Zoella Lykes* on September 26, 1944, at Port Hueneme, California, bound for Hawaii for further transport to the western Pacific to meet with their air assets and participate in the invasion of Yap. A change of orders, which the ship did not receive, directed them to Emirau Island to rendezvous with the air echelon to participate in the invasion of Leyte. Instead, *Zoella Lykes* continued to Hawaii, where – without orders – the captain attached his ship to a convoy headed for Ulithi and the ship dropped into a bureaucratic black hole.

The air detachment, which never knew the rest of the squadron was sitting in Pearl Harbor while it operated from MCAS Ewa, departed Hawaii on October 27 for an island-hopping trip to Emirau, arriving on October 27, 1944. With support from VMB-413 and -423, the aircrews were forced to fuel and service their own aircraft while Lt Col Sarles maintained the search for his missing air/groundcrew. VMB-611 made its combat debut on November 18, 1944, with a night heckling mission to Kavieng. During the next several weeks, the squadron followed the same routine as the others, with daily strikes against Kavieng and Rabaul.

Desperate Measures

Zoella Lykes arrived at Ulithi on November 5, 1944, and sat at anchor in a remote part of the harbor; squadron personnel were able to watch Task Force 38 come and go while food ran out, and they were forced to escape from the ship and steal supplies ashore.

Eventually, Lt Col Sarles discovered *Zoella Lykes* was still at Ulithi and, on December 24, 1944, he led a liberation mission to Falalop island in Ulithi, but unfortunately only the air element could be rescued. With 11 men crammed into each PBJ, the aircraft began their take-offs from the 3,200ft (975m) field on Falalap, with the tail hanging over the water at one end of the runway; pilots used

Underwing hardpoints for HVARs are visible on 306, a wingtip radar-equipped, late-model PBJ-1J of VMB-433.

every inch to "unstick" before meeting the sea at the other end, retracting the undercarriage with the props throwing up water spray as they flew low over the harbor to make their getaway.

Zoella Lykes continued its bizarre voyage with the groundcrew still aboard. The ship was finally directed to join a convoy headed for the Philippines in mid-January 1945, and finally arrived at Lingayen Gulf in Northern Luzon on February 7, 1945. Due to the lack of orders, the VMB-611 ground personnel did not leave the ship until February 24, when it was discovered the squadron was to be transferred to Mindoro at the other end of the Philippine Archipelago. Going aboard two LSTs (Landing Ship, Tanks), they made their way to Mindoro, where they arrived, on March 7, to discover the squadron was to be based at Zamboanga, Mindanao, where they finally disembarked on March 17, 1945, after a hellish trans-Pacific voyage of almost six months.

Modelers especially will find value in this photo of a 2,000lb (907kg) bomb in the fuselage bay of a VMB-433 PBJ Mitchell.

A sun-bleached PBJ-1D of VMB-433 flies over Emirau airfield. Note the two fixed .50 cal machine guns in the nose glazing and the early style gun packs scabbed on the fuselage.

In February 1945, Lt Col Sarles received word that VMB-611 would go to Mindanao to fly low-level strikes in support of American ground forces. In preparation, Sarles began sending his Mitchells out on low-level sweeps against New Ireland.

While the Emirau-based US Marines Mitchell squadrons continued the daily bombing missions against New Britain and New Ireland during early 1945, VMB-611 began its duty in the Philippines. Sarles led his squadron, by now equipped with new PBJ-1Js, to Zamboanga, and the unit was reunited on March 30, 1945. VMB-611 flew its first Philippine combat mission on April 5, 1945, with six PBJs attacking the Del Monte Airfield. Lieutenant Robert F. Jardes remembered the mission thus:

We dropped on the deck as we approached Del Monte, and in the finest traditions of aerial attack we came at the target out of the rising sun. I saw a truck barreling down the road to the airfield. Each airplane test-fired their nine forward-firing guns against this truck with a few hundred rounds of .50 caliber and when I flew over it as number six, it was burning fiercely. Over the field we spotted a white truck parked in the middle of the runway. Flying low down the runway, I spotted the fact it had no tires and was sitting on logs, a set-up for hidden anti-aircraft guns and warned the others. We proceeded to shoot up everything else on the field, leaving the white truck untouched. Finally, a bullet entered my open window and exited through the windshield, convincing me the Japs had the range and it was time to depart. When we returned to Zamboanga, we discovered all six airplanes had been holed multiple times, with one taking a hit in the starboard fin from a 90mm shell. Colonel Sarles questioned us about what we'd accomplished in return for this damage. After a long moment while we tried to remember what exactly had happened, one gunner spoke up: well, sir, we got this one truck for sure.

On August 10, 1945, captured Japanese Lt Minoru Wada, of the 100th Imperial Army Division, volunteered to lead VMB-611 to his division headquarters, where the bombers and US Marine Corps Corsairs destroyed the target.

Multi-Mission Success

Over a period of two months while Mindanao was secured, VMB-611 put in a stirring performance in supporting operations, striking shipping and other Japanese bases in the region to isolate enemy forces still fighting on the island. During April and May, VMB-611 flew 173 sorties, dropping approximately 245 tons of bombs, firing 800 rockets and almost a million rounds of machine gun ammunition. The Mitchells caused heavy casualties to the enemy, disrupted troop movements, rendered airfields inoperable, destroyed transportation and supplies, damaged artillery positions, and dented Japanese morale. The cost was nine dead, nine wounded and four aircraft lost.

On the morning of May 30, 1945, Lt Col Sarles led seven PBJs to sweep the Kibawe Trail. During the attack, he flew directly over concealed antiaircraft positions, and his Mitchell was hit in an engine, but one wing hit a tree as he tried to pull up. The PBJ-1J crashed into the ground and slid several hundred yards before it came to a halt. Some of the crew managed to scramble from the wreckage and made their separate ways to safety through Japanese lines.

For the men of VMB-611, Sarles' replacement, a man remembered by the originals of VMB-611 to this day as "Major Asshole," failed to fill the colonel's shoes, and morale suffered as a result. At one squadron reunion he attended in the 1970s, his reception was so cold the man beat a retreat after less than 30 minutes. In the two months left of the war, those originals not transferred put their efforts into surviving the poor leadership of men trying to get their "ticket punched" for postwar promotion, at the cost of many dead, due to what was seen by the combat crews as overarching incompetence.

Perhaps the strangest mission of all was that flown on August 10, 1945, five days before war's end. Japanese Lt Minoru Wada of the 100th Imperial Army Division had been captured, and he volunteered to lead the squadron to the division headquarters. Put aboard a PBJ, the graduate of the Japanese Army Military Academy was as good as his word, and the jungle headquarters was destroyed.

When the end of the war was announced, two of the few surviving original pilots still with the squadron – Capt Jardes and his copilot Lt LeMasters – went to the officers' mess and bought every bottle of whiskey they could, returning later to steal more. They took their treasure to the enlisted groundcrews and made sure the men finally had their only party since leaving Hawaii aboard *Zoella*

VMB-433 PBJ-1Ds make a low-level attack on a Japanese base. The weapons bays are open and "bombs away" has just been called.

Lykes. As Jardes put it: "I had a special affection for the mechanics, who had the hardest, dirtiest and most important job of all. If the guns don't work, you pull off and go home; if the bombs don't drop, you turn around and go home. But if the fans don't turn, you don't go anywhere."

As MAG-61's squadrons paid daily and nightly visits to Rabaul and Kavieng from Emirau, two new PBJ outfits commenced operations against the Japanese in the central Pacific, commencing in November 1944. The first was VMB-612, commanded by Lt Col Jack R. Cram, whose previous assignments included a tour as a PBY-5A Catalina pilot during the Guadalcanal campaign, in which he was awarded the Navy Cross and given the nickname "Mad Jack." Cram's VMB-612, which took the name "Cram's Rams," was one of just three PBJ squadrons trained to operate at night with radar-operated bombsights and search gear.

Three heavily weathered PBJ-1D Mitchells of VMB-433 go "feet dry" on their return to Emirau airfield.

VMB-612's PBJ-1Js arrived on Saipan on October 28, 1944, and Lt Col Cram led the first anti-shipping sweep to the Bonin Islands on November 13, 1944. Barren, rocky, volcanic islands, the Bonins were approximately 120 miles (193km) north of Iwo Jima and 800 miles (1,287km) from Saipan; Cram claimed a Japanese submarine and medium-sized freighter sunk during the strike. On November 15, he attacked three medium freighters and one larger vessel, as three hits were observed. During the last two weeks of November, the squadron lost two aircraft with three men killed.

In January and February 1945, VMB-612 made successful attacks on merchant shipping between Japan and the Bonin Islands. However, during one strike, 1st Lt Clifford James and his crew failed to return. While it was stationed on Saipan, VMB-612 flew 334 sorties and made 49 shipping attacks.

Iwo Jima-Bound

A forward echelon of six VMB-612 PBJ-Js and their crews moved to Iwo Jima on April 6, 1945, and the unit was placed under the control of VII Fighter Command. The southern shores of Japan were slightly more than 600 miles (965km) distant, well in reach of Cram's PBJs. The squadron conducted its first strike in Japanese home waters on April 10, attacking shipping with rockets in Kobe Harbor, and damaging a small merchant ship.

Four aircraft and crews were lost in three operational accidents during April, with another PBJ shot down by friendly fighters. VMB-612 was stationed on Iwo from April 10–July 28, 1945. During this time, the squadron flew 251 sorties with 83 targets located and 53 ships damaged or sunk. On July 13, 1945, segments of VMB-612 began arriving on Okinawa and were placed under the operational control of Fleet Air Wing 1, and they began anti-shipping sweeps along the northwestern coast of Kyushu. Their specially modified PBJ-1Js were capable of using the 11¾in (298mm) Tiny Tim rocket, with two of these stores being carried one on each side of the fuselage, above the bomb bay doors.

After free-fall launch, ignition was actuated by a lanyard that pulled free after the rocket had dropped clear of the aircraft. These weapons were used during nocturnal strikes against southern Japan during

This wonderful image shows a VMB-433 PBJ-1D performing a pre-take-off "mag" check at Emirau, as it waits for a Royal New Zealand Air Force Corsair to land.

the period between the Nagasaki Atom bomb drop on August 9, and Japanese surrender on August 15. Even with very few targets found during the closing days of the war, "Cram's Rams" flew 31 sorties and claimed 20 enemy ships damaged.

The last PBJ squadron to arrive in the Pacific Theater was VMB-613, commanded by Maj George W. Nevils, and it landed on Kwajalein in the Marshall Islands in early December 1944. On December 23, 1944, the squadron commenced missions against bypassed enemy islands in the Marshalls, and enemy shipping.

Ponape Island was the focus of attacks by VMB-613 on February 6, 1945, and the flak was intense. The third section came in at 1,500ft (457m), the lead Mitchell dropped its bombs and a moment later took a hit that killed the navigator instantly. The last aircraft took another hit in the right wing. Just as the PBJ came out of a wide turn and levelled out, the wing came off completely, and the aircraft crashed at the end of the runway, where it exploded on impact.

The Ponape mission was the highlight of a lackluster tour of duty for VMB-613. Afterwards, the squadron resumed flying anti-shipping missions without luck, since the central Pacific was devoid of enemy merchant shipping, and supplies for the bypassed Japanese garrisons were arriving by submarine. The squadron ended its days bombing these facilities in the Caroline and Marshall Islands.

At the end of the war in the summer of 1945, MAG-61's VMB-413, -423, -433, and -443 were transferred to Malabang, Mindanao, where they were joined by VMB-611. In October 1945, PBJs from VMB-611 escorted 1st MAW F4U-1D Corsair fighters and SBD-6 dive-bombers to Shanghai, China,

Pilot 1st Lt F. H. Kerns, Jr. poses with his PBJ-1D Mitchell, amusingly named *Fuzzy Fotos Inc.*, at Emirau, 1944.

for occupation duty and a show of strength. The MAG-61 PBJ squadrons were decommissioned at the end of November 1945.

VMB-613 then turned over its PBJs for disposal in October and was decommissioned on November 21. With the departure of VMB-611 to the US at the end of October, VMB-612 was the last PBJ squadron in the Pacific. On November 8, 1945, VMB-612 left Okinawa for the United States. The squadron was decommissioned on 14 March 1946, officially ending Marine Corps use of the PBJ Mitchell.

One PBJ-1H survives today; it was restored to original condition over the past ten years and is now operated by the Southern California Wing of the Commemorative Air Force, from its base at Camarillo Airport in Southern California.

APS-3 radomes adorn the starboard wingtips of these late-production PBJ-1J Mitchells, which also sport hardpoints for four High Velocity Aircraft Rocket stores under each wing.

Chapter 7
War's End
Betty's Dream

The dawn sky over Kyushu, southernmost island of the Japanese Home Islands, was clear on the morning of August 19, 1945, as Majs Jack McClure of 498th "Falcons" and Wendell Decker of 499th BS "Bats Outta Hell" orbited their B-25J Mitchell gunships offshore.

War in the Pacific had ended four days earlier, but the crews were alert for possible attacks by Japanese who refused their Emperor's decision to surrender. This was perhaps the most important rendezvous of the Pacific War as the two bombers waited. In an acknowledgment of their contribution to victory in this theater, these two B-25s of the 345th BG "Air Apaches" had been chosen to escort the official Japanese surrender delegation to the island of Ie Shima (Iejima), near Okinawa.

Navy aircraft of Task Force 58 stage a massive flyover of USS *Missouri* (BB-63) on September 2, 1945, following the signing of the surrender document that *Betty's Dream* had escorted to Japan on August 20. (NARA)

The white-painted Japanese G4M1 Betty bomber is watched by US personnel as it prepares to return to Japan, with the only existing copy of the surrender documents. (NARA)

Decker's top turret gunner spotted white dots to the north. The white dots soon resolved themselves as two G4M2 Betty bombers. Their camouflaged airframes had been overpainted hastily in white, with the well-known Hinomaru "meatball" insignia replaced by green crosses; four Mitsubishi A6M Zero escorts carried similar markings. As the B-25s swung in, the Zeros turned back.

Anticipation

At Ie Shima, air and groundcrews of the resident USAAF units, and any others who could arrange to get there, waited for the arrival of the four aircraft. As if staging a show for the victors, the two Bettys made a low pass over the runway before landing. Once on the ground, American military police speedily established a perimeter around the aircraft. The Japanese representatives were escorted to a waiting C-54 transport to fly on to Manila, Philippines, where they would meet with Gen Douglas MacArthur and other representatives of the victorious Allies to negotiate concrete details for the surrender of the Empire of Japan.

Once the C-54 was on its way, the MPs allowed the Americans on the field to surround the bombers; cameras clicked like insects as the men took photographs of the aircraft and their crews, who seemed overwhelmed by the friendly response they were receiving from those who they had been told would

Betty's Dream, the B-25J Mitchell gunship (44-30934) of the 499th BS, 345th BG "Bats Outta Hell" escorted the Japanese surrender delegation on return to Japan, with the only official copy of the surrender documents. The unit was given this honor in recognition of its role as the leading B-25 attack group in the Pacific over the previous two years. (NARA)

The surrender delegation boards a USAAF C-54 Skymaster transport to proceed to Manila. Once they had arrived at Ie Shima, the Japanese officers were flown to Manila, Philippines, where the formal surrender documents would be created. (NARA)